SOLUTIONS OF PROBLEMS IN THE EXERGY METHOD OF THERMAL PLANT ANALYSIS

by T J Kotas

A companion book to the textbook
The Exergy Method of Thermal Plant Analysis

Exergon Publishing Company UK Ltd
London 2012

Pages 288-317 published first in 1995 as Appendix G of the textbook The Exergy Method of Thermal Plant Analysis.
Pages 1-125 published in this volume first in 2012.

ISBN 978-1-78222-000-8

Book design, layout and production management by Into Print
www.intoprint.net
+44 (0)1604 832149

Printed and bound in UK and USA by Lightning Source

Preface to the Solution of the Problems

The problems solved in this companion book have been taken from Appendix G of the textbook. They correspond to the subject matter covered in the various chapters of the book. Since the Exergy Method is a relatively new area of Applied Thermodynamics, it was thought that the presentation of model solutions of problems of various types would be of some help both to teachers as well as self teaching students.

While solving the problems the advantages of the use of exergy analysis were demonstrated by pointing out and quantifying thermodynamic losses of various plant components and plant configurations. These were discussed at the end of the solutions under Comments. It is hoped that this will give the students a deeper understanding of the nature of irreversibilities of various kinds and their effect on plant performance.

Finally, I should like to express my sincere thanks to Mrs. Jean Hefford for her skilful typing of the text and the complex mathematical expressions.

Tadeusz J Kotas
Exergon Publishing co.uk.ltd
April 2012

Contents

Appendix G   Problems

General information

In the following problems, the data given below may be used, where necessary.

Standard gravitational acceleration	$g_E = 9.80$ m/s^2
Standard environmental pressure	$P^0 = 1.01325$ bar
Standard environmental temperature	$T^0 = 298.15$ K
Molar (universal) gas constant	$\tilde{R} = 8.3144$ kJ / kmol K

Composition of air for approximate calculations:

	Molar mass kg/kmol	Composition by volume	Composition by mass
N_2	28	0.79	0.767
O_2	32	0.21	0.233

Specific ideal gas constant $R = 0.287$ kJ/kg K
Molar mass $\tilde{m} = 28.96$ kg / kmol
Specific heat capacities (under perfect gas assumption)

$$c_p = 1.005 \text{ kJ/ kg K}$$
$$c_v = 0.718 \text{ kJ/kgK}$$
$$c_p/c_v = \gamma = 1.40$$

Thermodynamic properties of water, steam, ammonia and R-12 will be found tabulated with some other data in a simple and convenient form in a booklet compiled by Rogers and Mayhew (Reference 2.4, Appendix H).

In problems requiring properties of compressed liquids it is usually sufficiently accurate to assume that these are the same as those to be found in the saturated liquid state of the same temperature (i.e. the effect of pressure is assumed to be negligible).

Chapter 1 Review of the Fundamentals

1.1 A gas-fired central heating boiler is operating under steady state conditions. Give a list of four main forms of irreversibilities occurring during its operation.

1.2 A gearbox with a power input of 1000 kW operates steadily with an efficiency (power output/power input) of 0.96. The power lost through friction in the gearbox is transferred as heat to the environment at the temperature of T_0 = 300 K. Calculate the rate of entropy production associated with the operation of the gearbox.

1.3 Consider an adiabatic system consisting of a flywheel with a moment of inertia of 100 kgm^2 supported on bearings inside an evacuated box. Initially the system is at the environmental temperature of T_0 = 300 K and the flywheel spins at 60 rev/s but as a result of bearing friction, comes to rest and eventually the whole system reaches a uniform temperature. Assuming the heat capacity of the system to be 60 kJ/K, calculate:

(i) The final temperature of the system.
(ii) The entropy production of the system as the flywheel comes to rest.

Also, devise a method for restoring the system reversibly to its original state and calculate the necessary restoring work. Consider if the value calculated is compatible with the entropy production calculated under (ii).

1.4 A piece of steel with an isobaric heat capacity of 4 kJ/K cools naturally by exchanging heat with the environment from an initial temperature of 600 K to the final temperature of 300 K when it reaches a state of thermal equilibrium with the environment. Calculate the entropy produced in the cooling process.

1.5 A well insulated cylindrical copper rod of 1 m length is held at one end at 100 $^{\circ}$C whilst the other end is at 0 $^{\circ}$C. Calculate the entropy production rate per volume of the rod in W/(m^3 K). Take the thermal conductivity of copper as 385 W/(mK) and assume the conduction to take place steadily. If the rod had the form of a truncated cone, at which end of the rod would the entropy production rate per unit volume be greater, and why?

1.6 A hot stream (A) and a cold stream (B), both air, enter an adiabatic mixing chamber at the pressure of 1.02 bar. The mixed air emerges at the pressure of 0.99 bar. The mass flow rates and the temperatures of the entering streams are:

$$\dot{m}_A = 3 \text{ kg/s} \qquad \dot{m}_B = 2 \text{ kg/s}$$
$$T_A = 500 \text{ K} \qquad T_B = 400 \text{ K}$$

Assuming air to behave as a perfect gas and steady-state conditions calculate the entropy production rate of the mixing process.

1.7 A stream of ammonia enters a throttling valve at the rate of 0.1 kg/s at 20 ^oC in the saturated liquid state. It leaves at -10 ^oC as wet vapour. The specifc entropies and specific enthalpies of ammonia are given below:

State	Temp.	h/[kJ/kgK]	s/[kJ/kgK]
1 (inlet)	20 ^oC	275.1	1.044
2 (outlet)	-10 ^oC	304.1	1.185

The environmental temperature, T_0 is 300 K. Calculate:

(i) The heat transfer rate, $\dot{Q}_0$, between the ammonia stream and the environment.

(ii) The entropy production rate of the process.

1.8 Air is expanded in an adiabatic turbine from 3 bar to 1 bar whilst its temperature falls from 425 K to 340 K. Calculate the entropy production of the process per mass of the air. Assume air to behave as a perfect gas.

1.9 An adiabatic heat exchanger which acts as a mercury condenser and an H_2O evaporator is shown in Fig. G1. The heat tranfer rate is 500 MW. Using data given in the figure, calculate the entropy production rate of the heat transfer process. Assume absence of both pressure gradients and temperature gradients within the two fluids.

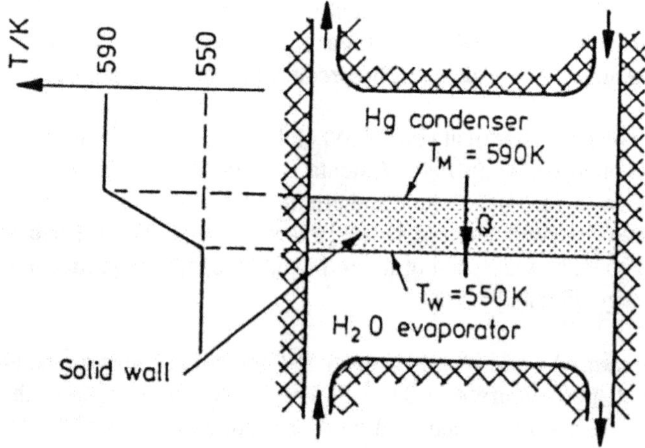

Fig. G1

1.10 Steam enters an adiabatic nozzle at 8 bar and 300 ^{o}C and leaves at 2 bar and 150 C. Calculate the entropy production of the expansion process in the nozzle per mass of the steam.

1.11 A feed water heater is to be constructed as a counter-flow heat exchanger. Steam, bled from the turbine, enters the heater in the superheated state and leaves as saturated liquid whilst feed water enters and leaves the heater as compressed liquid in the states given below:

Fluid	Position	$\frac{P}{\text{bar}}$	T	$\frac{h}{\text{kJ / kg}}$	$\frac{s}{\text{kJ / kgK}}$
Heating	At inlet	0.5	100 ^{o}C	2683	7.694
Steam	At exit	0.5	(81.3 ^{o}C)	340	1.091
Feed	At inlet	10	75 ^{o}C	313.9	1.015
Water	At exit	10	95 ^{o}C	398	1.250

(i) Calculate the ratio of mass flow rates of feed water to bled steam.

(ii) Calculate the entropy production per mass of the bled steam and hence, check if the assumed operating conditions of the feed heater do not violate the Second Law. Comment on the conclusion arrived at.

1.12 A fuel cell operates isothermally at 25 ^{o}C under steady state conditions using streams of methane (CH_4) and oxygen (O_2) at 25 ^{o}C and 1 atm in stoichiometric proportion. Liquid water and carbon dioxide gas exit at 25 ^{o}C and 1 atm. Neglecting kinetic and potential energy of the streams, calculate the maximum work per kmol of methane which the fuel cell could develop under ideal conditions. Use the following standard values of molar Gibbs functions of formation:

H_20, liquid - -237 180 kJ/kmol
CO_2, gas - -137 150 kJ/kmol
CH_4, gas - -50 790 kJ/kmol

1.13 Calculate the maximum work of the reaction

$$H_2(g) + \frac{1}{2}O_2(g) \rightarrow H_2O(g)$$

at 25°C and 1 atm. using the value of the equilibrium constant given by $\log_{10}K_p = 40.048$.

Check the calculated value, making use of the Gibbs function of formation of $H_2O(g)$ -228,590 kJ/kmol of H_2O.

Chapter 2 Basic Exergy Concepts

2.1 A cylindrical aluminium rod of 0.5 m length and 0.02 m diameter is held at one end at 100 °C and at the other at 0 °C. The rod is perfectly insulated and the thermal conductivity of aluminium is 203 W/(mK). Calculate the rate of heat transfer along the rod and the rates of thermal exergy flow at its two ends when the environmental temperature is 300 K.

2.2 An evaporator of a refrigeration plant refrigerates a cold chamber at the rate of 60 kW, maintaining its temperature at 250 K. Calculate the corresponding thermal exergy flow rate to the cold chamber when the environmental temperature is 300 K.

2.3 Calculate the specific exergy of water at 20 bar and 150 °C taking water vapour in the environment at the partial pressure of 0.01 bar and at 293.15 K as the reference substance. For these two states use:

	s/(kJ/kgK)	h/(kJ/kg)
State of water	6.331	1.840
State of vapour	2535.2	9.044

Show on a T - s diagram ideal reversible processes linking these two states.

2.4 Calculate the specific physical exergy of a stream of air for the following two states:

(i)	$P_1 = 3$ bar	$T_1 = 400$ K
(ii)	$P_2 = 0.8$ bar	$T_2 = 270$ K

when the environmental state is defined by $P_0 = 1$ bar and $T_0 = 290$ K. Assume air to be a perfect gas.
Calculate also the two components of physical exergy, $\varepsilon^{\Delta T}$ and $\varepsilon^{\Delta P}$ for the two cases.

2.5 Steam enters a well insulated heat exchanger at the rate of 0.1 kg/s as saturated vapour at 10 bar and leaves as saturated liquid whilst a stream of water enters at 20 bar and 20 °C and leaves at 165 °C. Calculate the rates of exergy transfer for the two streams as they pass through the heat exchanger. Assume negligible pressure drops in both streams and take $T_0 = 300$ K.

2.6 Steam is generated in a once-through boiler at a steady flow rate of 0.5 kg/s and at a constant pressure of 50 bar from an initial temperature of 30 ^{0}C to the final temperature of 400 ^{0}C. Calculate the change in the exergy flow rate of the steam between the inlet and outlet when the environmental temperature is 300 K. Use the following values of specific enthalpy and entropy:-

State	$\frac{h}{kJ/kg}$	$\frac{s}{kJ/kgK}$
Initial state	125.7	0.436
Final state	3196	6.646

2.7 Calculate the standard molar chemical exergies of O_2, CO_2 and H_2O_{vap}. Take their standard mole fractions in the environment to be:-

O_2 - 0.02040 CO_2 - 0.000294 H_2O_{vap} - 0.0088.

Using these values, calculate the standard molar chemical exergy of a mixture of these gases with the following molar composition:

O_2 - 0.52 CO_2 - 0.45 H_2O_{vap} - 0.03.

The standard environmental temperature is 298.15 K.

2.8 Calculate the molar chemical exergy of methane gas (CH_4). The molar Gibbs functions of formation for CH_4, CO_2 and H_2O_{vap} are -50810 kJ/kmol, -394390 kJ/kmol and -228590 kJ/kmol respectively. Use the values of molar chemical exergy of CO_2 and H_2O_{vap} and O_2 calculated in Problem 2.7.

2.9 Calculate the molar chemical exergy of a mixture of atmospheric air and methane in the proportion of 20 to 1 by mole. Take the molar chemical exergy of methane 836504 kJ/kmol and $T_0 = 298.15$ K.

2.10 Calculate the standard chemical exergy of CO from the following set of data:-

$$\Delta \tilde{g}^{o}_{fco} = -137160 \text{ kJ/kmol}$$

$$\tilde{\varepsilon}^{o}_{02} = 3975 \text{ kJ/kmol}$$

$$\tilde{\varepsilon}^{o}_{c} = 410530 \text{ kJ/kmol}$$

2.11 Calculate the standard molar chemical exergy of the gaseous products of combustion of composition given in the table below.

Gaseous constituent	Composition by volume in per cent	Standard molar chemical exergy kJ / kmol
CO	4.09	275350
N_2	69.30	720
H_2O	20.48	11770
CO_2	6.13	20170

2.12 Calculate the standard molar chemical exergy of the gaseous fuel of the composition given in the table below.

Gaseous constituent	Composition by volume in per cent	Standard molar chemical exergy kJ / kmol
H_2	47.0	238350
CH_4	41.0	836510
C_2H_4	2.5	1366610
CO	7.5	275350
N_2	2.0	720

2.13 Using an ideal model, show that the molar exergy of a mixture of perfect gases made up of common constituents of the atmosphere can be expressed as follows:-

$$\tilde{\varepsilon} = \sum_{i=1}^{i=N} x_1 \left\{ \tilde{c}_{pi}(T_1 - T_0) - T_0 \ln \frac{T_1}{T_0} + \tilde{R}T_0 \ln \frac{P_1 x_1}{P_0 x_{i00}} \right\}$$

where

x_i - mole fraction of the i-th constituent of the mixture.

x_{i00} - mole fraction in the atmosphere of the i-th constituent of the mixture.

P_1, T_1 - pressure and temperature of the mixture.

P_0, T_0 - pressure and temperature of the atmosphere.

$\tilde{R}$ - molar (universal) ideal gas constant.

$\tilde{c}_{pi}$ - molar specific heat capacity of the i-th constituent of the mixture.

N - number of constituents of the mixture.

Exhaust gases at a pressure of 1 bar, a temperature of 250 ^{o}C are discharged with negligible velocity at the rate of 0.12 kmol/s from an internal combustion engine into the atmosphere. Using data given in the table, calculate the exergy flow rate of the gas stream. The environmental temperature is 25 ^{o}C. The constituents of the mixture may be assumed to behave as perfect gases.

Constituents of the exhaust gases	Molar composition of the products x_i	Molar composition of the atmosphere x_i	Mean molar heat capacity $\tilde{c}_{pi}$ /[kJ / kmolK]
CO_2	0.05	0.0003	40.0
H_20	0.12	0.0177	34.0
O_2	0.08	0.2060	29.9
N_2	0.75	0.7760	29.2

2.14 A stream of a combustible mixture specified in Problem. 2.9 is entering a combustion chamber at the rate of 0.5 kg/s. The pressure, temperature and velocity of the stream are 1.2 bar, 100 ^{o}C and 100 m/s respectively, whilst the environmental parameters are P_0 = 1 bar and T_0 = 25 ^{o}C. Calculate the exergy flow rate of the stream. Use the following values for the components of the mixture, assuming perfect gas behaviour.

	CH_4	Air
Molar mass $\tilde{m}$ /[kg / kmol]	16.042	28.96
Sp. heat capacity c_p /[kJ/kgK]	2.334	1.006

2.15 Calculate the specific non-flow exergy of air for the two states specified in Problem 2.4.

2.16 A pressure vessel of 0.1 m^3 capacity contains dry air at 5 bar and 400 K. Assuming perfect gas behaviour, calculate the exergy of the air contained in the vessel. Take T_o = 300 K, P_o = 1 bar for the environment and γ = 1.4 for air. Also, calculate the exergy of the system defined by the inside surface of the pressure vessel when it is completely evacuated.

2.17 Treating air and water vapour as perfect gases, show that molar exergy of a stream of humid air can be expressed as follows:-

$$\tilde{\varepsilon}_1 = \frac{\tilde{c}_{p,a} + \tilde{\omega}_1 \tilde{c}_{p,v}}{1+\tilde{\omega}_1} T_o \left(\frac{T_1}{T_o} - 1 - \ln \frac{T_1}{T_o} \right) + \tilde{R} T_o \ln \frac{P_1}{P_o}$$

$$+ \tilde{R} T_o \left(\ln \frac{1+\tilde{\omega}_{oo}}{1+\tilde{\omega}_1} + \frac{\tilde{\omega}_1}{1+\tilde{\omega}_1} \ln \frac{\tilde{\omega}_1}{\tilde{\omega}_{oo}} \right)$$

where the subcripts v and a refer to vapour and air respectively, and $\tilde{\omega}_1$ and $\tilde{\omega}_{oo}$ are the mole fraction ratios (x_v / x_a) of the humid air in the initial state and in the environment respectively.

Chapter 3 Elements of Plant Analysis

3.1 A rigid, well-insulated, pressure vessel of 0.1 m^3 capacity is fully evacuated. A valve which isolates the vessel from the atmosphere is then opened until the air pressure in the vessel becomes equal to that of the outside atmosphere, P_0 = 1 bar. Calculate the irreversibility which took place during the filling process.

3.2 A pressure vessel of 0.1 m^3 capacity contains carbon dioxide CO_2, at 30 bar and 10 °C. As a result of fracture of the vessel the gas escapes into the atmosphere which is at a pressure of 1 bar and temperature of 10 °C. The mole fraction of CO_2 in the atmosphere is 0.0003. Calculate the destruction of exergy caused by the escape of CO_2.

3.3 A frictionless piston and cylinder assembly contains 0.0025 kg of air initially at 1 bar and 20 °C. The air then undergoes a polytropic compression process specified by $PV^{1.25}$ = const. to a final pressure of 4 bar, during which heat is transferred to the environment at T_0 = 293.15 K. Calculate the irreversibility of the process. Assume air to behave as a perfect gas.

3.4 A steady stream of Refrigerant 12 enters a throttling valve at the rate of 0.1 kg/s at the pressure of 5.673 bar in the saturated liquid state. It leaves at the pressure of 2.191 bar and a dryness fraction of 0.21. Taking the temperature of the environment to be 300 K, calculate:

(i) The heat-transfer rate between the valve and the environment.
(ii) The irreversibility rate of the process.

Also, calculate the irreversibility rate of the process under *adiabatic* conditions and comment on the results obtained.

Use the following values of properties of Refrigerant 12:

State	$\frac{P}{\text{bar}}$	$\frac{h_f}{\text{kJ / kg}}$	$\frac{s_f}{\text{kJ / kgK}}$	$\frac{h_g}{\text{kJ / kg}}$	$\frac{s_g}{\text{kJ / kgK}}$
1	5.673	54.87	0.2078	-	-
2	2.191	26.87	0.1080	183.19	0.7020

3.5 Exhaust steam from the turbine of a back-pressure steam plant is used in a heat exchanger to heat a stream of air for a warm-air central heating system of a building. The steam enters the heat exchanger at a rate of 0.5 kg/s in the dry saturated state at a pressure of 0.20 bar and condenses there at constant pressure leaving in the saturated liquid state. The air flow rate through the heat exchanger is 37.5 kg/s, the pressure and temperature at inlet being 1.01 bar and 20 °C, and those at outlet 0.99 bar and 50 °C respectively. Calculate:

(i) Heat transfer rate to the environment.
(ii) The total irreversibility rate associated with the heat exchange process.

Assume air to behave as a perfect gas. Temperature of the environment is 280 K.

3.6 Saturated steam at a pressure of 4 bar is fed to a mixing type of feed-water heater. Feed water is pumped from a condenser, which it leaves at a pressure of 0.05 bar and in the saturated liquid state, into the feed-water heater. The combined streams leave the heater as saturated liquid at 4 bar. Assume the process in the feed pump to be reversible and adiabatic.

(a) How much steam enters the heater per mass of feed water entering?
(b) Find the irreversibility of the process per mass of the feed water entering the heater. Take $T_o = 283$ K.

3.7 Calculate the exergy flow rates of the three air streams involved in the mixing process in Problem1.6 and hence calculate its irreversibility rate. Compare this with the value obtainable from the Gouy-Stodola relation. Take the environmental temperature $T_o = 300$ K and pressure $P_o = 1$ bar.

3.8 A pipe carries a stream of brine at the rate of 0.5 kg/s. The brine temperature at the inlet to the pipe is 240 K, but owing to heat transfer from the environment, the temperature at the pipe outlet is 243 K. Neglecting pressure losses due to viscous friction, calculate the irreversibility rate of the process when the temperature of the environment is 300 K. Take for brine $c_p = 2.85$ kJ/kgK.

Represent the irreversibility rate of the process as an area on a *T-S* diagram.

3.9 Steam enters a de-superheater at 400 ^oC and 20 bar and at the rate of 0.5 kg/s. Calculate:

(i) the necessary mass flow rate of water in the saturated liquid state at 20 bar which must be used to reduce the steam to dry saturated state.
(ii) the irreversibility rate of the process. Take $T_o = 300$ K.

3.10 In a vapour-compression refrigerator the evaporator is in the form of a counter-flow heat exchanger in which ammonia, the working fluid, exchanges heat with brine. In a test run the following values of the parameters were obtained:

Ammonia mass flow rate	$\dot{m}_a = 0.45$ kg/s
Brine mass flow rate	$\dot{m}_b = 16$ kg/s
Ammonia pressure	$P_a = 1.902$ bar
Ammonia dryness fraction at inlet	$x_{a1} = 0.25$
Ammonia superheat at outlet	$\Delta T_{su} = 10$ K

Brine temperature at inlet $T_{b1} = -5\ ^{o}C$

Brine temperature at outlet $T_{b2} = -15\ ^{o}C$

Assuming pressure losses and changes in kinetic and potential energy to be negligible, calculate:

(i) Heat transfer rate between the evaporator and the surroundings.
(ii) Irreversibility rate of the heat transfer process.

Assume for brine a constant value of isobaric specific heat capacity of 2.85 kJ/kgK. Take the temperature of the environment as 20 ^{o}C.

3.11 With reference to the heat exchanger described in Problem 1.9, calculate the thermal exergy flow rates for the heat transfer from the mercury at T_M and to the water substance at T_W.. Hence, calculate the irreversibility rate of the heat transfer process. Comparing this with the value obtainable from the Gouy-Stodola relation. Take the environmental temperature $T_o = 300$ K.

3.12 State what is the input and output, expressed in terms of exergy, for a vapour-compression refrigerator. Hence, formulate the rational efficiency of this plant. Show how an alternative expression for the rational efficiency of the vapour-compression refrigerator can be formulated with the aid of the exergy balance.

A refrigeration plant, using ammonia as the working fluid, operates between pressures of 11.67 bar and 2.908 bar. The ammonia leaves the condenser in the saturated liquid state and is then expanded through an adiabatic valve to the evaporator pressure. The fluid leaves the evaporator in saturated vapour state and is then compressed in an adiabatic compressor, with an isentropic efficiency of 0.73, to the condenser pressure. The heat rejected in the condenser is dissipated in the environment which is at a temperature of 25 ^{o}C. The evaporator heat exchanger is placed in a cold-room in which the temperature remains constant at -6 ^{o}C. Calculate:

(i) the specific irreversibilities and the corresponding efficiency defects of the four principal components, and
(ii) the rational efficiency of the plant.

Also, construct an exergy balance for the plant in the form of a pie diagram.

3.13 In the condenser of an ammonia heat pump (shown diagramatically in Figure. G2), a stream of air having a mass flow of 0.5 kg/s, is heated at a steady rate. Ammonia enters the heat exchanger at a pressure of 15.54 bar and a temperature of 90 ^{o}C and condenses at constant pressure to the saturated liquid state. The air enters the heat exchanger at a pressure of 1 bar and a temperature of 16 ^{o}C. Its pressure increases over the axial fan to 1.005 bar and then drops as a result of friction back to 1 bar at the exit from the heat exchanger. The exit air temperature is 30 °C. The overall efficiency of the motor is 0.7 and the isentropic efficiency of the axial fan is 0.8. The condenser may be assumed to be adiabatic whilst the losses of the electric motor are dissipated in the surroundings. Calculate:

(i) the electric power input to the fan,
(ii) the mass flow rate of the ammonia stream, and
(iii) the rational efficiency of the fan - condenser system defined by the control surface shown in the diagram, taking the isobaric exergy increase of the air stream as the output.

Assume air to be a perfect gas. The temperature of the environment is 280 K.

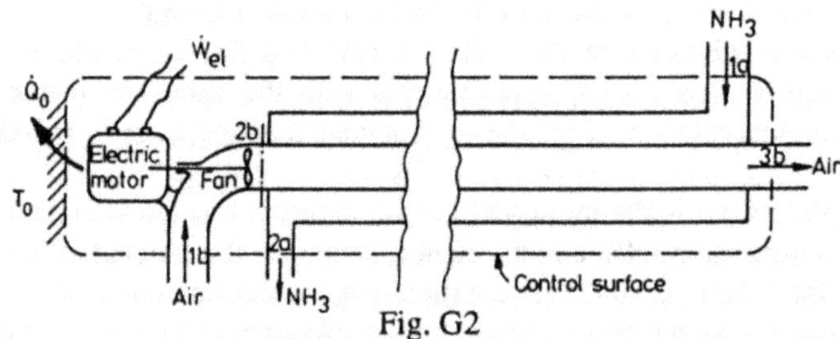

Fig. G2

3.14 An experimental domestic heat pump takes in dry atmospheric air at a pressure of 0.98 bar and a temperature of 5 °C and compresses it in an adiabatic compressor to a pressure of 2 bar. The compressed air is then passed through a heat exchanger inside the house where it is cooled. At the exit from the heat exchanger, the pressure and temperature of the air are 1.96 bar and 32 °C, respectively. The air is then expanded in an adiabatic turbine to the atmospheric pressure and discharged back to the atmosphere. The turbine and the compressor are mechanically coupled and the net power input is provided from an external source of mechanical power. The temperature inside the house is 20 °C. Assuming air to be a perfect gas and taking the isentropic efficiency of both the turbine and the compressor to be 0.80, calculate:

(i) specific irreversibilities of the three principal plant components.
(ii) external irreversibility due to mixing of the discharged air with the ambient air.
(iii) the rational efficiency of the plant.

3.15 A single stage reciprocating air compressor with an aftercooler (see Figure G3), operates with the following temperatures and pressures:

T_o = 288 K	$P_1 = P_o$ = 1 bar
T_2 = 400 K	P_2 = 4 bar
T_3 = 293 K	P_3 = 3.90 bar

where the subscripts 1, 2 and 3 refer to the states of the air stream indicated in Figure G3 and 0 denotes the environmental state. The mass flow rate of the air is 0.1 kg/s and the compressor operates with an overall isothermal efficiency of 0.68. Calculate:

(i) the necessary shaft power and the heat-transfer rates from the compressor, $\dot{Q}_c$ and from the aftercooler, $\dot{Q}_a$.

(ii) the exergy terms necessary for the construction of exergy balances for the two sub-regions shown in Figure G3, detailing separately, in the case of sub-region B, the component of irreversibility due to pressures losses, $\dot{I}_B^{\Delta P}$, and that due to heat transfer over a finite temperature difference $\dot{I}_B^{\Delta T}$. Hence, construct a Grassmann diagram for the two-subregion system.

Assume air to be a perfect gas.

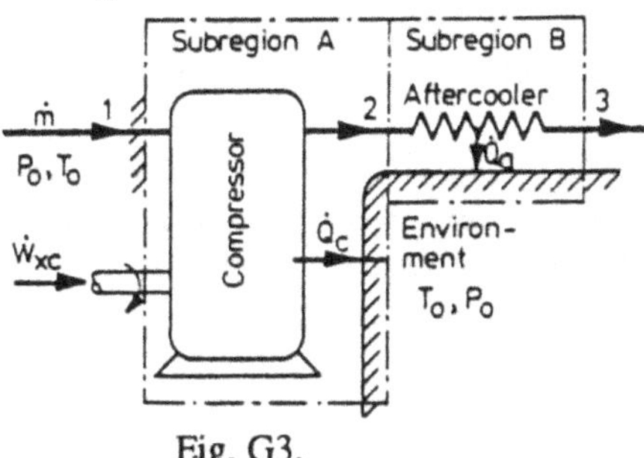

Fig. G3.

3.16 A cold chamber is maintained at a steady temperature of 260 K using a vapour-compression refrigerator with a rational efficiency ψ = 0.4, whilst the environmental temperature is 300 K. The chamber is illuminated by means of a 500 W incandescent bulb. What is the total irreversibility rate attributable to this source of illumination?

What will this irreversibility rate be if the temperature of the cold chamber is reduced to 240 K the other parameters remaining unchanged. Comment on the results obtained.

3.17 An inventor submits to the Patent Office a proposal for an engine which, he claims, delivers a net power of 720 kW when using a fuel having an enthalpy of combustion of -40000 kJ/kg at the rate of 0.025 kg/s. In the combustion chamber the maximum temperature is limited to 1000 K whilst the engine uses a heat sink which is maintained at 200 K by a refrigerator driven by the engine. The refrigerator uses the atmosphere at 300 K as its heat sink. Check whether the inventor's claims can be justified.

3.18 An engineer is to design an adiabatic steadily operating device which would take a stream of 0.1 kg/s of dry atmospheric air at 1 atm and 25 °C and, utilising shaft power input from an external source, produce from it two streams of air, both at atmospheric pressure, but one at 75 °C and the other at 0 °C. Calculate the minimum shaft power input required to perform this task and the corresponding mass flow rates of the output streams. Assume air to be a perfect gas.

3.19 In a closed-type regenerative feed water heater steam bled from a turbine, enters at 6 bar as saturated vapour and condenses there leaving as saturated liquid. Feed water enters the feed pump at 0.05 bar as saturated liquid. On leaving the feed pump, it is heated in the feed heater to a temperature 7 K below the saturation temperature of the bled steam. There is a heat transfer from the feed heater to the environment which amounts to 3 per cent of the enthalpy change of the bled steam in the feed heater. Neglecting pressure losses and the feed-pump work, calculate:

(i) the fraction of steam bled from the turbine for feed heating, and
(ii) the irreversibility taking place in the feed heater per unit mass of the steam entering the turbine. The temperature of the environment is 290 K.

3.20 In a surface heat exchanger heat transfer takes place at a steady rate $\dot{Q}$ between a condensing fluid at temperature T_1 and an evaporating fluid at tempereature T_2 with negligible pressure losses. Show that the irreversibility rate for this type of heat transfer process can be expressed by

$$\dot{I} = \dot{Q}\frac{T_o \Delta T}{T_M^2}$$

where $\Delta T = T_1 - T_2$
and $T_M = \sqrt{T_1 T_2}$

Chapter 4 Exergy analysis of simple processes

4.1 Calculate the irreversibilities per mass of the refrigerants involved in the throttling process of a refrigeration plant when the refrigerant used is

(i) ammonia,
(ii) Refrigerant 12

The initial state of the process is the saturated liquid state at 10 ^{o}C and the final state is at the evaporator pressure corresponding to the saturation temperature of -20 ^{o}C. Assume the process to be adiabatic and the environmental temperature 300 K.

Calculate now the irreversibility rates of the two throttling processes on the basis of the same refrigeration duty of 100 kW assuming that the refrigerant leaves the evaporator in the dry saturated state and pressure losses in the heat transfer processes are negligible. Comment on the results obtained. Use the following values of properties of the two refrigerants

	Initial State T_{SAT}= 10 ^{o}C		Final State T_{SAT}= -20 ^{o}C			
	$\frac{h_f}{kJ/kg}$	$\frac{s_f}{kJ/kgK}$	$\frac{h_f}{kJ/kg}$	$\frac{s_f}{kJ/kgK}$	$\frac{h_g}{kJ/kg}$	$\frac{s_g}{kJ/kgK}$
NH_3	227.8	0.881	89.8	0.368	1420.0	5.623
R-12	45.37	0.1752	17.82	0.0731	178.73	0.7087

4.2 Using data given for the nozzle described in Problem 1.10, calculate:

(i) steam exit velocity
(ii) irreversibility of the expansion process in the nozzle per mass of the steam
(iii) the rational efficiency of the expansion process.

Assume negligible steam velocity at inlet to the nozzle and an environmental temperature of 293 K.

4.3 Air is expanded in an adiabatic turbine from 3 bar to 1 bar whilst its temperature falls from 425 K to 340 K. Neglecting changes of kinetic and potential exergies, calculate:

(i) irreversibility of the expansion process in the turbine per mass of the air,
(ii) isentropic efficiency, η_{isen}, of the process
(iii) rational efficiency, ψ, of the process.

Comment on the relative magnitudes of η_{isen} and ψ.

Assume air to behave as a perfect gas and an environmental temperature of 293 K.

4.4 The power output of an adiabatic, steadily operating steam turbine is controlled by throttling the steam before it enters the turbine. The steam is supplied from a boiler at 15 bar and 350 ^oC and the turbine exhausts into a condenser in which pressure is maintained at 0.05 bar.

Assuming a turbine isentropic efficiency of 0.8, calculate:

(i) for the case when the turbine is operating at the full supply pressure:

(a) specific irreversibility of the expansion process in the turbine
(b) the specific work output
(c) the rational efficiency of the turbine

(ii) for the case when the steam supply pressure is throttled to 10 bar:

(a) specific irreversibilities of the throttling process and the expansion process in the turbine.
(b) the specific work output
(c) the rational efficiency of the turbine, and of the combined valve-turbine sub-system.

The environmental temperature is 290 K.

4.5 Using the $\varepsilon - h$ chart for air in Fig. E.2, estimate the specific irreversibility of adiabatic throttling (isenthalpic) processes from 8 MPa to 0.1 MPa for the cases, when the initial state is at the temperature of (a) 140 K and (b) 120 K. Comment on the relevance of these values to air liquefaction processes.

4.6 In a steam operated ejector (Fig. G.4)the motive steam leaves the nozzle at the rate of 1.2 kg/s with a velocity of 1100 m/s, a pressure of 0.091 bar and a dryness fraction of 0.85. In the mixing chamber, the motive steam mixes completely with dry saturated steam entering at right angles to the axis of the ejector at the rate of 0.6 kg/s with a velocity of 150 m/s. Mixing takes place adiabatically at constant pressure and with conservation of axial momentum.

Determine the state of the steam leaving the mixing chamber and the irreversibility rate of the mixing process. The environmental temperature is 293 K.

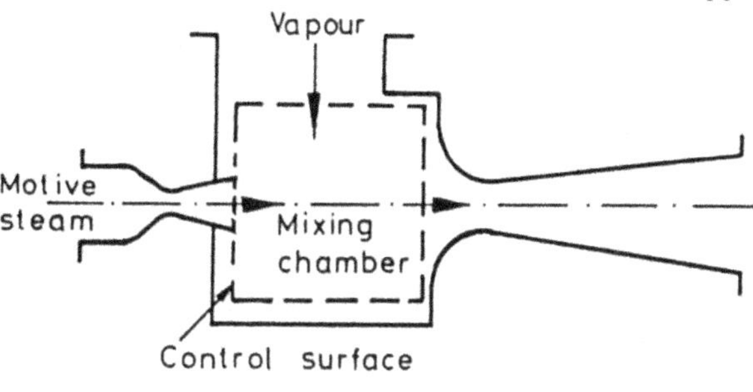

Fig. G.4

4.7 Two compressors operating in series are used to deliver compressed air to an industrial process at the steady rate of 0.3 kg/s. The first compressor takes in atmospheric air at 1 bar and 300 K and compresses it to 3 bar whereupon the air is transferred to the second compressor where it is further compressed to the delivery pressure of 9 bar. Both compressors may be assumed to be adiabatic and to operate with the same isentropic efficiency of 0.75. Neglecting changes of kinetic and potential exergies between the inlets and the outlets of the compressors, calculate for each compressor:

(i) the irreversibility rate
(ii) the net power input
(iii) the rational efficiency.

Assume air to behave as a perfect gas.

4.8 The compressed air plant described in Problem 4.7 has been modified by introducing between the two compressors an intercooler which reduces the temperature of the air before the entry to the second compressor to the ambient value. How will the results obtained in Problem 4.7 be affected by this modification?

4.9 In an adiabatic turbine steam expands from state (1) with an isentropic efficiency, η_s . The final state (2) lies in the wet steam region at a pressure P_c for which the saturation temperature is T_c. Show that the irreversibility rate of the process can be expressed as

$$\dot{I} = \dot{m}(h_1 - h_2)\left(\frac{1}{\eta_s} - 1\right)\frac{T_o}{T_c}$$

4.10 State what is the input and output, expressed in terms of exergy, for a heat exchanger involving two streams of which one has negligible pressure losses. Hence, formulate an exergetic efficiency for this heat exchanger. Show how an alternative expression for the exergetic efficiency can be formulated with the aid of the exergy balance.

Exhaust steam from the back-pressure steam plant is used in a counterflow heat exchanger to heat a stream of air for a warm-air central heating system of a building. The steam enters the heat exchanger at a rate of 0.4 kg/s in the dry saturated state at a pressure of 0.20 bar and condenses there, leaving in the liquid state at the same pressure and a temperature of 41.5 ^oC. The pressure and temperature of the air stream at inlet are 1.02 bar and 10 ^oC and those at outlet are 0.99 bar and 50 ^oC respectively. Neglecting changes in kinetic and potential energy of the fluids and heat transfer between the heat exchanger and its surroundings, calculate:

(i) the mass flow rate of air,
(ii) the irreversibility rate
(iii) the rational efficiency of the process.

Assume air to be a perfect gas and the temperature of the environment 10 ^oC.

4.11 In a water chilling plant water enters a heat exchanger at a temperature of 25 ^oC and leaves at 5 ^oC· Ammonia (NH_3) enters the heat exchanger as a mixture of liquid and vapour with a dryness fraction of 0.18 at a pressure of 3.983 bar and leaves as saturated liquid. The mass flow rate of water is 1 kg/s and that of ammonia 0.083 kg/s. Assuming the heat exchanger to operate with negligible pressure losses and changes in kinetic and potential energy, calculate:

(i) the irreversibility rate of the process,
(ii) the rational efficiency.

The environmental temperature is 300 K.

4.12 In an air pre-heater, a stream of air is heated by a stream of products of combustion in counter flow. In a test run, the following operating parameters were obtained:

	Mass flow rate	Inlet		Outlet	
	$\dot{m}$/[kg/s]	Temp.T/K	Press. P/bar	Temp. T/K	Press. P/bar
Products of combustion	0.50	403	1.05	310	1.00
Air	0.55	283	1.10	358	1.05

Calculate:

(i) Heat transfer rate to the surroundings.
(ii) Irreversibility rate of the heat exchange process.
(iii) The rational efficiency of the heat exchanger.

Assume both gases to be perfect. Take for products of combustion, $c_p = 1.15$ kJ/kgK, $\gamma = 1.333$, and for air $c_p = 1.0$ kJ/kgK, $\gamma = 1.4$, and the temperature of the surroundings $T_0 = 290$ K.

4.13 Assuming that the heat exchanger described in Problem 4.12 could be made to operate with negligible heat loss and pressure losses and with near-zero temperature difference at one end, what would be the (intrinsic) irreversibility rate of the process and the corresponding rational efficiency. The mass flow rates of the gases and the initial temperatures of the two gas streams are to be taken the same as in Problem 4.12. Comment on the results obtained.

4.14 A mixture consisting of 90% heptane (C_7H_{16}) and 10% octane (C_8H_{18}) by mole is made at a steady rate from the pure constituents at 25 ^oC and 1 atm. Calculate:

(i) standard molar chemical exergy of the mixture,
(ii) irreversibility of the mixing process per mole of the mixture when $T_0 = 25$ ^oC.

Take the standard molar chemical exergy of heptane and octane 4787300 kJ/kmol and 5440030 kJ/kmol, respectively.

Assume that the mixture is ideal.

4.15 A stream of air enters a gas separation plant at a steady rate of 1.0 mol/s at atmospheric pressure and temperature, $P_o = 1$ bar and $T_o = 293$ K. The air is separated in the plant into two streams. One of the streams, with a molar flow rate of 0.2 mol/s consists of 90% oxygen and 10% nitrogen by volume. The two streams leave the plant at atmospheric pressure and temperature. Assuming atmospheric air consists only of oxygen and nitrogen with the mole fractions O_2 - 0.21, N_2 - 0.79, calculate:

(i) The molar composition of the other stream leaving the gas separation plant.
(ii) the minimum exergy input necessary to sustain the given rate of gas separation in the plant. Compare this result with values which can be obtained from the $\varepsilon - x$ chart for air given in Fig. E.4 (p. 284).

4.16 Carbon monoxide (CO) and water vapour enter an adiabatic reaction chamber as separate streams at equal steady molar flow rates and both at 150 ^oC and

1 atm. The reaction is complete and the products exit as a mixture of hydrogen (H_2) and carbon dioxide (CO_2) at 1 atm. Neglecting changes in kinetic and potential exergy, calculate:

(i) the temperature of the product stream,
(ii) the irreversibility of the process per kmole of CO.

Use data from Appendix A and D. The temperature of the environment is 25 °C.

4.17 Methane (CH_4) enters an adiabatic combustion chamber and burns completely with 120% of stoichiometric air entering in a separate stream. Both the reactants are supplied at 1 atm. and 25 °C and the products leave at 1 atm. with negligible changes in kinetic and potential exergy. Calculate:

(i) the exit tempereature of the products
(ii) the irreversibility of the reaction per kmol of the fuel.

Use data from Appendix A and D. The temperature of the environment is 25 °C.

4.18 The combustion process described in Problem 4.17 is modified by pre-heating the fuel (CH_4) to 80 °C and the air to 200 °C. Calculate the exit temperature of the products and the irreversibility of the process per kmol of CH_4. Comment on the values obtained in the two combustion processes.

4.19 Because of certain technological and environmental considerations, it has been decided to limit the exit temperature of the products in the combustion process with pre-heating described in Problem 4.18 to T_3 = 1800 °C. This should be achieved by increasing the air-fuel ratio. Calculate:

(i) the necessary air excess over the stoichiometric requirement
(ii) the irreversibility of the combustion process per kmol of the fuel.

Comment on the merits of this process.

4.20 Calculate the exit temperature of the products and the process irreversibility per kmol of carbon monoxide (CO) when the latter is burnt completely under adiabatic conditions and constant pressure with:

(i) 130% of the stoichiometric amount of air,
(ii) 200% of the stoichiometric amount of air.

The reactants enter the combustion chamber separately at the environmental temperature of 25 °C and the pressure of 1 atm.

Comment on the effect of air excess on the process irreversibility.

4.21 Consider an idealized model of a steady-state process for producing quicklime (CaO) from calcium carbonate ($CaCO_3$) according to the following decomposition reaction:

$$CaCO_3 \rightarrow CaO + CO_2$$

In this process $CaCO_3$ is fed to the reactor at 25 °C and the products are removed at 200 °C, all at the common pressure of 1 atm. The necessary heating is provided to the reactor at a constant temperature T_r. Calculate the irreversibility of the process in the reactor space per kmol of $CaCO_3$ for the case when (a) T_r = 1250 K and (b) T_r = 1450 K. Assume the reactor to be well thermally insulated and take the temperature of the environment 25 °C. For quicklime at 200 °C, the following values may be used:

$$\tilde{c}_p^h = 47.60 \text{ kJ / kmol K and } \tilde{c}_p^s = 45.59 \text{ kJ / kmol K.}$$

Comment on the relationship between the temperature T_r and the process irreversibility.

Chapter 5 Examples of Thermal and Chemical Plant Analysis

5.1 A coal-fired, steam power plant is designed to deliver 5000 kW of electric power. Superheated steam leaves the boiler at 15 bar and 200 °C and is expanded in an adiabatic turbine to the condenser pressure of 0.05 bar. The steam is subsequently condensed to saturated liquid, which is then pumped back to the boiler. Divide the plant into the following subregions:

(i) boiler,
(ii) turbine,
(iii) electric generator
(iv) condenser with the cooling tower,
(v) feed pump

and construct a Grassmann diagram for the plant assuming the exergy rate of the feed water to have zero value. Also calculate the overall rational efficiency of the plant.

The chemical exergy of the coal is 1.06 times its NCV, the combustion efficiency of the boiler is 0.80, the turbine isentropic efficiency is 0.70 and the generator electrical efficiency is 0.90.

Assume that the feed-pump power input and the associated irreversibility can be neglected. The temperature of the environment is 10 °C.

5.2 A refrigerator-heat pump plant is to be installed at a new leisure centre to provide both refrigeration to an ice rink and heating to a swimming pool. It is estimated that the refrigeration duty necessary to keep the ice at -4 °C is 50 kW whilst the required water temperature of the swimming pool is 28 °C. The NH_3 plant selected will operate on a simple vapour-compression cycle between the pressures of 2.680 bar in the evaporator and 13.89 bar in the condenser. The compressor may be assumed to be adiabatic with an isentropic efficiency of 0.82. It is driven by an electric motor with an efficiency of 0.9. The throttling process starts from the saturated liquid state and the compression process starts from the dry saturated vapour state. Pressure losses in the heat exchangers and stray heat transfer may be considered negligible. Calculate the rational efficiency of the plant and construct a Grassmann diagram for it, assuming zero value for the exergy flow rate for the NH_3 stream at the inlet to the compressor.

5.3 A vapour-compression ammonia refrigerator, driven by an electric motor, includes a counter-flow heat exchanger in which saturated liquid leaving the condenser is subcooled by saturated vapour leaving the evaporator. The pressure in the condenser is 10.34 bar and the NH_3 leaving it is cooled to 16 °C in the heat exchanger and then throttled in to the evaporator which operates at 2.908 bar. After being superheated in the heat exchanger the NH_3

vapour is compressed adiabatically, with an isentropic efficiency of 0.8, to the condenser pressure.

The refrigerator delivers a refrigeration duty of 30 kW in a cold chamber which it maintains at -6 °C whilst the environmental temperature is 20 °C. The efficiency of the electric motor is 0.9. To facilitate calculations associated with the heat exchanger, the specific heat capacities of the liquid and the vapour may be taken as 4.75 kJ/kgK and 2.35 kJ/kgK respectively.

Calculate the irreversibility rates of the six main plant components and construct a Grassmann diagram for the plant assuming that the exergy flow rate of the ammonia vapour at the entry to the compressor has zero value. Also, calculate the rational efficiency for the plant when operating as described above and for the case when it is designed to operate as above but without the heat exchanger. Compare these values and comment.

Pressure losses in heat exchange and stray heat transfers are to be neglected.

5.4 Figure G.5 is a plant schematic and Fig. G.6 shows the thermodynamic processes in *T-s* coordinates for the Heylandt nitrogen liquefaction plant. The plant is shown sub-divided into seven sub-regions numbered using roman numerals. Gaseous nitrogen is delivered to the compressor at the rate of 0.1 kg/s at the environmental pressure and temperature $P_0 = 1.013$ bar and $T_0 = 300$ K (point O) where it is compressed with an isothermal efficiency of 0.7 to $P_1 = 202.6$ bar and temperature $T_1 = T_0$. A fraction F = 0.55 of the nitrogen stream

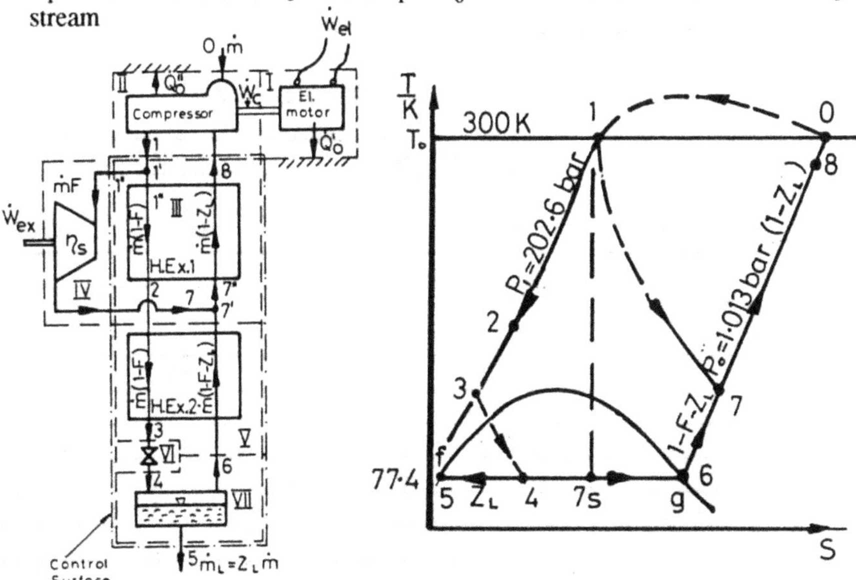

Fig. G.5

Fig. G.6

is diverted to an adiabatic expander (IV), and after expansion to P_0, the stream mixes *reversibly* with the return stream, in state 7, between the two heat exchangers, (III and V). The forward stream undergoes a throttling process (3-4) which results in wet vapour. The separation of the two phases takes place reversibly under gravity (VII).

Other operating parameters

Expander isentropic efficiency	-	0.75
Electric motor efficiency	-	0.95
Compressor mechanical efficiency	-	0.88
Expander mechanical efficiency	-	0.90
Return stream final temperature	T_8 =	297 K
Ideal gas constant of N_2	R =	0.297 kJ/kgK

Principal assumptions

(i) Pressure losses in the heat exchangers are negligible.
(ii) Heat transfer between the apparatus and the environment is negligible.
(iii) Kinetic and potential components of exergy of the stream are negligible.

Construct a Grassmann diagram for the plant using physical exergy values. Sketch a $\tau - \dot{H}$ diagram using calculated values of τ for states 1, 2, 3, 6, 7 and 8, and approximate curves joining these points. Also, calculate the rational efficiency for the plant taking the exergy of liquid nitrogen and the expander power as the total plant exergy output. Compare your results with those given in the book for the simple Linde plant and the Linde plant with auxiliary refrigeration and discuss reasons for the differences.

Thermodynamic properties of nitrogen

	P = 1.013 bar		P = 206.6 bar	
$\frac{T}{\text{K}}$	$\frac{h}{\text{kJ/kg}}$	$\frac{s}{\text{kJ/kgK}}$	$\frac{h}{\text{kJ/kg}}$	$\frac{s}{\text{kJ/kgK}}$
77.4 sat.liq.	29.4	0.418		
77.4 sat.vap.	228.7	2.994		
100	253.0	3.270	87.2	0.784
110	263.6	3.371	106.3	0.966
120	274.2	3.463	124.5	1.124
260	420.5	4.272	376.9	2.548
280	441.3	4.349	404.8	2.652
300	462.1	4.421	431.5	2.744

Datum: $h = 0$ and $s = 0$ for saturated liquid at 63.15 K.
Abstracted from Technical Note 129, January 1962,
National Bureau of Standards.

5.5 In an oil-fired boiler, process steam is generated at a constant pressure and a steady rate. Feed water is delivered to the plant at 150 bar and 25 ^{0}C and the steam leaves at 500 ^{0}C. The fuel oil used has the composition by mass C-87% and H-13%, (NCV) = 42500 kJ/kg and c_p = 2.02 kJ/kgK. The fuel and the air are delivered to the plant at the standard environmental pressure and temperature and are then pre-heated in series using the products of combustion as shown in Fig. G.7. Air excess of 30% over the stoichiometric requirement is provided. Assuming no stray heat transfer and no pressure losses, calculate:

(i) the temperature, T_{F2}, to which the fuel is pre-heated, given that the temperature difference between the products at the hot end of the fuel preheater (sub-region A) is 20 K,

(ii) the temperature of the products, T_{G4}, at the point of discharge to the atmosphere,

(iii) the adiabatic flame temperature, T_{G1}

(iv) irreversibilities per 1 kg of the fuel for the five sub-regions indicated in the figure,

(v) the rational efficiency of the plant.

Also, construct a Grassmann diagram for the plant.

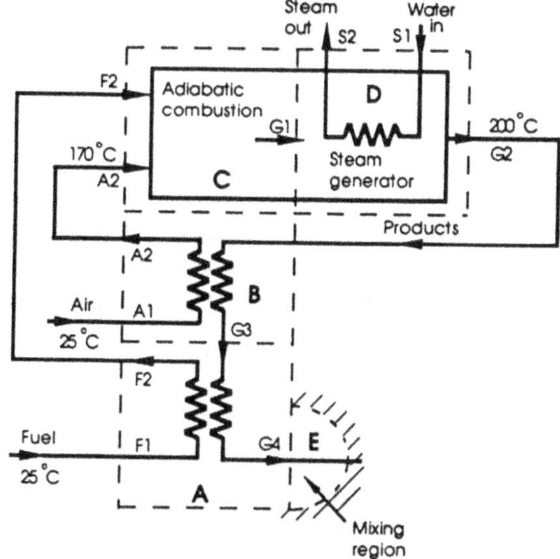

Fig. G.7

Chapter 6 Thermoeconomic Applications of Exergy

6.1 The total cost of a new power plant to be built is to be financed by taking out a loan of £12 x 10^6. The loan company charges 15% annual interest rate compounded monthly. It is estimated that £180,000 will be available monthly from the income for loan repayment. How many years will it take to repay the loan.

6.2 Exergy flow rate of steam delivered by a boiler is 2 x 10^4 kW, whilst the fuel cost rate is 500 £/h. If the capital cost of the boiler is £4.5 x 10^6, its period of operation per year is 7000h and the capital-recovery factor may be taken as 0.1868 y^{-1}, what is the minimum unit cost of exergy of the steam delivered?

6.3 The operation of the refrigeration plant analysed in Problem 3.12 is altered by reducing the evaporator pressure from 2.908 bar to 2.680 bar $\left(\theta_{SAT} = -12\ ^{o}C\right)$, other operating parameters remaining unchanged. Assuming that the plant is to deliver a refrigeration duty of 100 kW under both sets of operating conditions, calculate:

(i) irreversibility rates of the four principal components for both sets of operating conditions.

(ii) changes in the irreversibility rates in the four principal plant components as fractions of the change in the irreversibility rate in the whole plant resulting from the change in the evaporator pressure. Comment on these results.

(iii) The coefficient of structural bonds for the evaporator related to the change in its operating pressure:

$$\sigma_{EV,P} = \left(\frac{\Delta \dot{I}_{TOT}}{\Delta \dot{I}_{EV}}\right)_{P_{ev}=\text{var}}$$

6.4 The evaporator of the refrigerator considered in Problems 3.12 and 6.3 is to be optimised using the structural method with the annual cost of owning and operating the plant as the objective function. The capital cost of the heat exchanger can be expressed as

$$C_D^C = K_B + \frac{K_A Q}{U \Delta T}$$

where A - heat transfer area

Q - heat transfer rate

U - overall heat transfer coefficient (taken to be constant)

ΔT - temperature difference in heat transfer

K_A, K_B - constants

Using the following data:

t_{op}	=	4000 h/y
a^{c}	=	0.1868 y^{-1}
K_A	=	200 £/m^2
K_B	=	£460
U	=	2 kW/m^2K
c_{IN}^{ε}	=	0.05 £/kWh
$\sigma_{D_1 \Delta T}$	=	1.98 (see Problem 6.3)
T_O	=	288 K
T_M	=	266 K as first approximation (see Problem 3.20)

Using the structural method, calculate the optimum temperature difference ΔT, the corresponding irreversibility rate, heat transfer area and the capital cost of the heat exchanger for a heat transfer rate of 100 kW.

6.5 The electric power plant described in Problem 5.1 is to be upgraded by using a turbine with an isentropic efficiency of 0.75. Calculate the new values of irreversibility rates of the sub-regions of the plant assuming unchanged power output, and compare them with those calculated in Problem 5. Obtain changes in the irreversibility rates of the sub-regions as a percentage of the change in the irreversibility rate of the plant as a hole. Hence, calculate the coefficient of structural bonds for the turbine.

$$\sigma_{t,\eta} = \left(\frac{\Delta \dot{I}_{TOT}}{\Delta \dot{I}_t} \right)_{\eta=var}$$

6.6 The steam turbine in the power plant considered in Problems 5.1 and 6.5 is to be optimised using the structural method with the annual cost of owning and operating the plant as the objective function. The capital cost of the turbine of the required capacity can be expressed as a function of the isentropic effliciency, η_S, by the following empirical formula:

$$C_t^c \,/\, £ = -2.44 \text{ x } 10^6 + 10.15 \text{ x } 10^6 \, \eta_s{}^{1.8}$$

for $0.6 < \eta_s < 0.9$.

Also, use the expression for the irreversibility rate of a steam turbine given in Problem 4.9 and the following data:

a^{C}	=	0.124 y^{-1}
t_{op}	=	6000 h/y
NCV	=	25.0 MJ/kg, (for the coal)
c_F^{SP}	=	£65/tonne

$\sigma_{t,\eta}$ = 4.47 (see Problem 6.5)

Calculate the optimum isentropic efficiency and the corresponding capital cost of the turbine.

6.7 The total cost of the plant described in Problem 5.2 is to be financed by taking out a loan of £64,000 at the fixed annual interest rate of 12 per cent and is to be repaid in equal monthly instalments over the projected life of the plant of 20 years. It is anticipated that the plant will operate for 7800 hours each year and the unit cost of electric energy will be 0.07 £/kWh.

Using the *global approach* (see p. 230), and assuming that both products of the plant are of equal importance, calculate:

(i) the unit cost of *exergy* delivered by the plant to the ice rink and the swimming pool,
(ii) the corresponding unit cost of *energy* delivered to the swimming pool.

Neglect the effect of maintenance, repairs, etc. on the unit costs.

6.8 The swimming pool referred to in Problems 5.2 and 6.7 is to be provided with auxiliary heating from a gas-fired water heater with a nominal output of 60 kW and a thermal efficiency of 0.78 based on the net calorific value of the gas of 50014 kJ/kg. The chemical exergy of the fuel is 52145 kJ/kg. The capital cost of £6,000 of the water heater, which includes the cost of its installation, will be financed on the same terms as those of the refrigerator - heat pump. Assuming the cost of the energy of the fuel based on its net calorific value, to be 0.014 £/kWh, calculate:

(i) the rational efficiency of the plant,
(ii) the unit costs based on:
  (a) the *energy*, and
  (b) the *exergy*

delivered to the swimming pool.

6.9 The approximate thermoeconomic modelling method described in Section 6.8, p. 231, has been used to model a series of vapour-compression ammonia refrigerators of the same type and the same structure. The following constants appearing in expressions (6.84), (6.87) and (6.88) have been determined from an exact optimisation:

C_0^C = £4400, k = 1850 £/kW, m = 2.3

(i) Calculate the optimum value of the rational efficiency, ψ_{OPT}, and the corresponding capital cost, C^{C} for a plant with a refrigeration duty of 100 kW when the cold chamber temperature is 250 K and that of the environment 293 K. The following thermoeconomic parameters are assumed:

Annual interest rate	$i = 0.12\ y^{-1}$
Period of operation of the plant per year	$t_{op} = 5000$ h/year
Period of repayment of the loan (equal to the plant life)	$N_y = 20$ years
Unit cost of input (electrical) exergy	$c^{\varepsilon}_{IN} = 0.07$ £/kWh

(ii) Investigate the effect of variation in the annual interest rate, i, and the unit cost of electricty, c^{ε}_{IN} on ψ_{OPT} and C^{C} for the following values:

i = 0.08 and 0.16 per year
c^{ε}_{IN} = 0.05 and 0.09 £/kWh

whilst keeping the remaining parameters constant as in (i), above.

Comment on your results.

Answers

Chapter 1

1.2 0.133 kW/K

1.3 (i) 418.4 K,(ii) 19 96 kJ/K, 5988 kJ

1.4 1.227 kJ/K

1.5 37.78 W/(m^3K)

1.6 0.072 kW/K

1.7 2.9 kW, 4.4 x 10^{-3} kW/K

1.8 0.091 kJ/kg K

1.9 61.6 kW/K

1.10 0.047 kJ/kgK

1.11 (i) 27.8, (ii) -0.07 kJ/kg

1.12 560720 kJ/kmol of CH_4

1.13 228595 kJ/kmol of H_2

Chapter 2

2.1 12.75 W, 2.499 W, -1.253 W

2.2 12 kW

2.3 209.8 kJ/kg

2.4 (i) 108.26 kJ/kg, 16.82 kJ/kg, 91.44 kJ/kg
(ii) -17.86 kJ/kg 0.72 kJ/kg, -18.58 kJ/kg

2.5 -68.06 kW, 34.31 kW

2.6 603.65 kW

2.7 3940 kJ/kmol, 20158 kJ/kmol, 11733 kJ/kmol, 9477 kJ/lmol

2.8 836504 kJ/kmol

2.9 39358.9 kJ/kmol

2.10 275357.5 kJ/kmol

2.11 13224.5 kJ/kmol

2.12 507135 kJ/kmol

2.13 323.5 kW

2.14 703.5 kW

2.15 (i) 31.7 kJ/kg, (ii) 1.52 kJ/kg

2.16 26.32 kJ, 10 kJ

Chapter 3

3.1 10 kJ

3.2 3164 kJ

3.3 0.0133 kJ

3.4 0.273 kW, 0.1979 kW

3.5 (i) 54.0 kW, (ii) 146.05 kW

3.6 0.219, 49.2 kJ/K

3.7 145.4 kW, 30.8 kW, 154.5 kW, 21.6 kW

3.8 1.03 kW

3.9 0.119 kg/s, 21.1 kW

3.10 3.27 kW, 19.6 kW

3.11 245.8 MW, -227.3 MW, 18.5 MW

3.12 58.44 kJ/kg, 51.1 kJ/kg, 15.8 kJ/kg, 18.9 kJ/kg 0.214, 0.187, 0.058, 0.069, 0.472

3.13 (i) 0.37 kW,
(ii) 5.42 x 10^{-3} kg/s,
(iii) 0.337

3.14 (i) 12.55 kJ/kg, 7.155 kJ/kg, 11.99 kg
(ii) 0.535 kJ/kg (iii) 0.0758

3.15 (i) 16.85 kW 5.60 kW, 10.75 kW
(ii) $\dot{E}_2 = 13.21$ kW
$\dot{E}_3 = 11.25$ kW

3.16 692.3 W, 812.5 W

3.17 $\dot{W}_{MAX} = 700$ kW

3.18 0.2086 kW, 0.064 kg/s, 0.036 kg/s

3.19 (i) 0.248, (ii) 69.5 kJ/kg

Chapter 4

4.1 (i) 9.65 kJ/kg, (ii) 1.976 kJ/kg,
(i) 0.809 kW, (ii) 1.48 kW

4.2 (i) 757.6 m/s, (ii) 13.77 kJ/kg
(iii) 0.954

4.3 (i) 26.68 kJ/kg, (ii) 0.742,
(iii) 0.762

4.4 (i) 190 kJ/kg, 782 kJ/kg, 0.805
(ii) 52.8 kJ/kg, 175.5 kJ/kg,
742 kJ/kg, 0.809, 0.765

4.5 (a) 153 kJ/kg, (b) 92 kJ/kg

4.6 0.96, 733.3 m/s 229.6 kW

4.7 (i) 7.781 kW, 7.781 kW, (ii) 44.47 kW, 66.33 kW,
(iii) 0.825, 0.883

4.8 (i) 7.781 kW, 7.781 kW, (ii) 44.47 kW, 44.47 kW
(iii) 0.825, 0.825

4.10 (i) 23.5 kg/s, (ii) 137.9 kW
(iii) 0.307

4.11 (i) 5.7 kW, (ii) 0.38

4.12 (i) 12.22 kW, (ii) 9.136 kW,
(iii) 0.29

4.13 0.464 kW, 0.954

4.14 (i) 4850867 kJ/kmol,
806 kJ/kmol

4.15 (i) 0.0375, 0.9625,
(ii) 781.96 W

4.16 (i) 658 ^{o}C, 9925 kJ/kmol

4.17 (i) 1804 ^{o}C,
(ii) 259900 kJ/kmol

4.18 1925 ^{o}C, 219975 kJ/kmol

4.19 (i) 0.32, (ii) 222395 kJ/kmol

4.20 (i) 2018 ^{o}C, 58432 kJ/kmol
(ii) 1494 ^{o}C, 96180.8 kJ/kmol

4.21 4577 kJ/kmol, 10944 kJ/kmol

Chapter 5

5.1 0.152

5.2 0.402

5.3 0.693 kW, 0.979 kW,
1.58 kW, 0.102 kW,
0.176 kW, 0.501 kW,
0.420, 0.444

5.4 0.445

5.5 48.8 ^{o}C, 1850 ^{o}C, 66.5 ^{o}C,
4.2 kJ,
212.4 kJ, 14461 kJ,
12062 kJ, 76.3 kJ, 0.41

Chapter 6

6.1 12 years

6.2 3.1 p/kWh

6.3 5.616 kW 4.915 kW,
1.602 kW, 2.563 kW, 1.98,

6.4 3.32 K, 1.35 kW,
15.06 m^2, £3472

6.5 4.41

6.6 0.80, £4.35 x 10^6

6.7 0.35 £/kWh, 0.0093 £/kWh

6.8 (i) 0.02, (ii) (a) 0.0196 £/kWh
(b) 0.738 £/kWh

6.9 (i) 0.463, £27025

Solutions of the Problems

Problem 1.1

The main forms of irreversibility are due to:

(i) exothermic chemical reaction,
(ii) heat transfer over temperature differences,
(iii) viscous friction in the gases and the water,
(iv) mixing of gas and air.

Problem 1.2

Transmissions efficiency

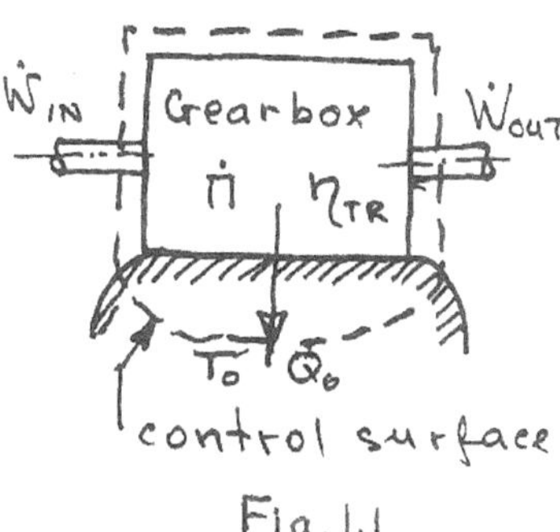

Fig.1.1

$$\eta_{TR} = \frac{\dot{W}_{OUT}}{\dot{W}_{IN}}$$

From the energy balance

$$\dot{Q}_o = \dot{W}_{IN} - \dot{W}_{OUT}$$

Hence,

$$\dot{Q}_o = \dot{W}_{IN}(1 - \eta_{TR}) = 1000(1 - 0.96) = 40 \text{ kW}$$

Steady-state, entropy production rate in an open system

$$\dot{\Pi} = (\dot{S}_{OUT} - \dot{S}_{IN}) - \sum \frac{\dot{Q}_i}{T_i}$$

However, here

$$\dot{S}_{OUT} = 0, \ \dot{S}_{IN} = 0$$

Also, $\dot{Q}_i = -\dot{Q}_o$ and $T_i = T_o = 300 \text{ K}$

Thus $\dot{\Pi} = \frac{\dot{Q}_o}{T_o} = \frac{40}{300} = \underline{0.133 \text{ kW/K}}$

Problem 1.3

For an adiabatic closed system at constant pressure, with C = const

$$\Pi = (S_2 - S_1) - C \ln \frac{T_f}{T_o}$$

The initial kinetic energy is transformed into an increase of the internal energy of the system.

$$\therefore \mathbf{E}_{KIN} = C(T_f - T_o)$$

But $\mathbf{E}_{KIN} = \dfrac{1}{2} I w^2 \quad \therefore T_f = T_o + \dfrac{I w^2}{2C}$

Q = 0

Fig. 1.3.a

Data: $\quad$ I = 100 kg.m^2 $\quad$ N = 60 rev/s

$\quad\quad$ C = 60 kJ/K $\quad$ T_o = 300 K

(i) $\therefore T_f = 300\,\text{K} + \dfrac{100[\text{kg.m}^2](2\pi \times 60)^2[\text{K}] \times 10^{-2}[\text{kJ}]}{60[\text{kJ}][\text{s}^2][\text{J}]}$

$= \underline{418.4\ \text{K}}$

(ii) Entropy production

$$\Pi = 60 \ln \frac{418.4}{300} = \underline{19.96\ \text{kJ/K}}$$

Restoring work

The system is cooled reversibly as shown.

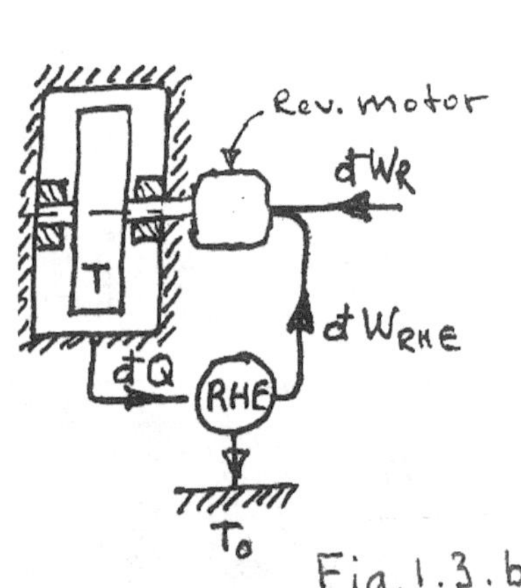

Fig. 1.3.b

$$\mathrm{d}W_{RHE} = \frac{T - T_o}{T} \mathrm{d}Q$$

$$\therefore W_{RHE} = \int_{T_f}^{T} \frac{T - T_o}{T} \mathrm{d}Q \qquad \mathrm{d}Q = C\mathrm{d}T$$

$$= C\left[(T_f - T_o) - T \ln \frac{T_f}{T_o}\right]$$

$\therefore W_R = E_{KIN} - W_{RHF} = C T_o \ln \dfrac{T_f}{T_o} = T_o \Pi$ (see above)

$= \underline{5988\ \text{kJ}}$ $\therefore$ The answer is compatible with that for (ii).

Problem 1.4

Entropy change of the combined system = entropy produced,

i.e. $\quad \Pi = \Delta S_{\mathrm{STEEL}} + \Delta S_{\mathrm{ENV}}$

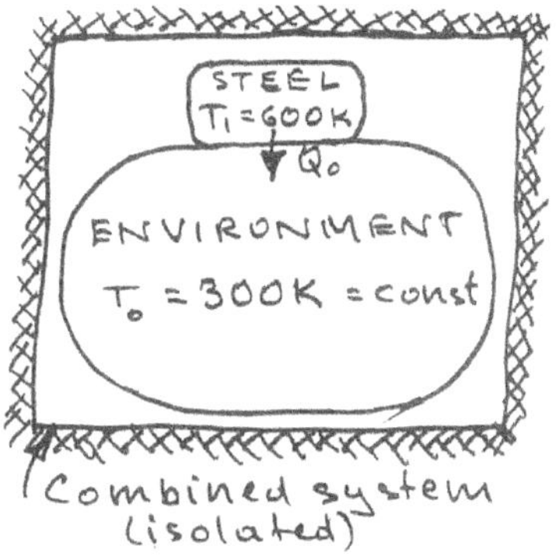

At $\quad P = \mathrm{const}$
and $\quad C_s = \mathrm{const}$

$$\Delta S_{\mathrm{STEEL}} = C_s \ln\frac{T_o}{T_1} \qquad C_s = 4\ \mathrm{kJ/K}$$

$$\Delta S_{\mathrm{ENV}} = \frac{Q_o}{T_o} \qquad \text{where } Q_o = C_s\,(T_1 - T_o)$$

Fig. 1.4

substituting

$$\Pi = 4\,[\mathrm{kJ/K}]\ln\frac{300}{600} + \frac{4\,[\mathrm{kJ/K}](600-300)\ \mathrm{K}}{300\ \mathrm{K}}$$

$$= (-2.773 + 4)\ \mathrm{kJ/K} = \underline{1.227\ \mathrm{kJ/K}}$$

Problem 1.5

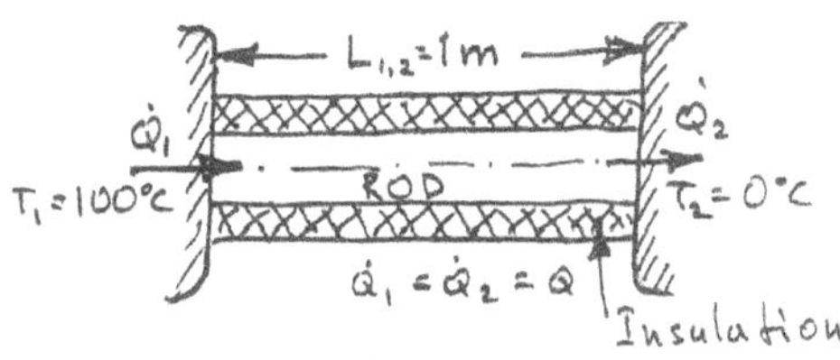

Fig. 1.5

$$k = 385\ \mathrm{W/(mK)}$$

Entropy production rate

$$\dot{\Pi} = \left(\dot{S}_{\mathrm{OUT}} - \dot{S}_{\mathrm{IN}}\right) - \sum_i \frac{Q_i}{T_i} = \frac{\dot{Q}_2}{T_2} - \frac{\dot{Q}_1}{T_1} = \dot{Q}\frac{T_1 - T_2}{T_1 T_2} \qquad \text{(a)}$$

For steady-state conduction

$$\dot{Q} = -kA\frac{\mathrm{d}T}{\mathrm{d}x} \qquad \text{(b)}$$

In the case of an adiabatic rod with a constant cross-sectional area A

$$\frac{\mathrm{d}T}{\mathrm{d}x} = \frac{T_2 - T_1}{L_{1,2}} \qquad \text{(c)}$$

Also, the volume of the rod is given by

$$V = AL_{1,2} \qquad \text{(d)}$$

Combining (a) → (d)

$$\frac{\dot{\Pi}}{V} = \frac{k}{L_{1.2}^2} \times \frac{(T_2 - T_1)^2}{T_1 T_2} - \frac{385[\text{W/mK}]100^2[\text{K}^2]}{1^2[\text{m}^2] \times 373.15 \times 273.15[\text{K}^2]}$$

$$= \underline{37.78\ \text{W/(m}^3\text{K)}}$$

<u>A rod in the form of a truncated cone</u>

From (b), with $\dot{Q} = \text{const}$

$$\frac{\text{d}T}{\text{d}x} \propto \frac{1}{A} \tag{e}$$

As follows from (e), the temperature gradient, and hence also $\dot{\Pi}/V$ would be greater at the smaller end of the rod.

Problem 1.6

Energy equation with $\Delta KE = 0 \quad \Delta PE = 0$

$$\dot{H}_\text{A} + \dot{H}_\text{B} = \dot{H}_\text{C} \tag{a}$$

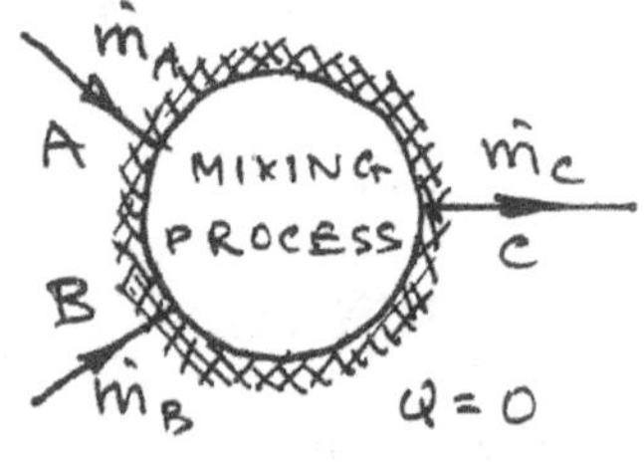

Fig. 1.6

Conservation of mass

$$\dot{m}_\text{A} + \dot{m}_\text{B} = \dot{m}_\text{C} \tag{b}$$

for a perfect gas, from (a) and (b)

$$\dot{m}_\text{A} c_P (T_\text{A} - T_\text{o}) + \dot{m}_\text{B} c_P (T_\text{B} - T_\text{o}) = (\dot{m}_\text{A} + \dot{m}_\text{B}) c_P (T_\text{C} - T_\text{o})$$

Hence,
$$T_\text{c} = \frac{\dot{m}_\text{A} T_\text{A} + \dot{m}_\text{B} T_\text{B}}{\dot{m}_\text{A} + \dot{m}_\text{B}} \tag{c}$$

Given that: $\dot{m}_\text{A} = 3\ \text{kg/s}$ $\quad \dot{m}_\text{B} = 2\ \text{kg/s}$
$T_\text{A} = 500\ \text{K}$ $\quad T_\text{B} = 400\ \text{K}$

Substituting in (c)

$$T_\text{c} = 460\ \text{K}$$

Entropy production rate, (with $\dot{Q}_\text{o} = 0$)

$$\begin{aligned}\dot{\Pi} &= \dot{S}_\text{OUT} - \dot{S}_\text{IN} \\ &= (\dot{m}_\text{A} + \dot{m}_\text{B}) s_c - \dot{m}_\text{A} s_\text{A} - \dot{m}_\text{B} s_\text{C} \\ &= \dot{m}_\text{A}(s_\text{C} - s_\text{A}) + \dot{m}_\text{B}(s_\text{C} - s_\text{B})\end{aligned}$$

For a perfect gas

$$\dot{\Pi} = \dot{m}_A c_P\left[\ln\frac{T_C}{T_A} + \frac{\gamma-1}{\gamma}\ln\frac{P_2}{P_1}\right] + \dot{m}_B c_P\left[\ln\frac{T_C}{T_B} - \frac{\gamma-1}{\gamma}\ln\frac{P_2}{P_1}\right]$$

$$= 3\times1.005\left[\ln\frac{460}{500} - \frac{0.4}{1.4}\ln\frac{0.99}{1.02}\right]$$

$$+ 2\times1.005\left[\ln\frac{460}{400} - \frac{0.4}{1.4}\ln\frac{0.99}{1.02}\right]$$

$= \underline{0.072 \text{ kW/K}}$

Comment: Entropy production in this process is due to: (a) mixing of two streams of air in different thermodynamic states, and (b) viscous friction, which manifests itself as a pressure drop between the inlets and the outlet of the mixing chamber, from 1.02 bar to 0.99 bar.

Problem 1.7

(i) The energy balance - steady flow, with $\dot{W}_x = 0,\ \Delta KE = 0\ \ \Delta PE = 0$ is

$$\dot{Q}_o = \dot{m}(h_2 - h_1)$$
$$= 0.1(304.1 - 275.1)$$
$$= \underline{2.9 \text{ kW}}$$

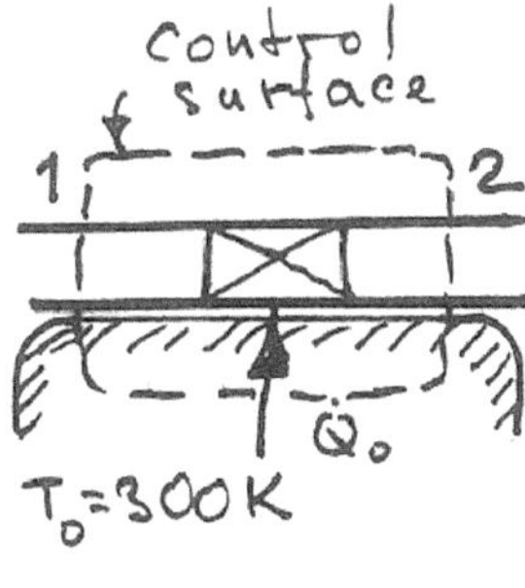

Fig.1.7

(ii) Entropy production rate, steady flow

$$\dot{\Pi} = \dot{m}(S_2 - S_1) - \frac{\dot{Q}_o}{T_o}$$

$$= 0.1\ (1.185 - 1.044) - \frac{2.9}{300} = \underline{0.0044 \text{ kW/K}}.$$

Problem 1.8

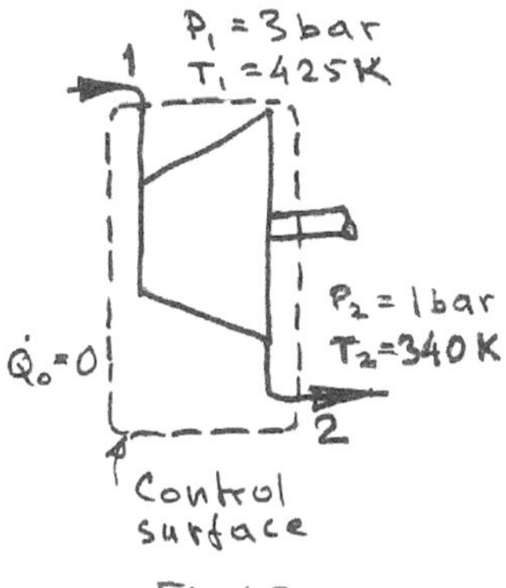

Fig.1.8

For an adiabatic open system undergoing a steady flow process

$$\dot{\Pi} = \dot{S}_{OUT} - \dot{S}_{IN}$$

Per mass of the air

$$\pi = \dot{\Pi} / \dot{m} = s_2 - s_1$$

assuming perfect gas behaviour

$$\pi = c_P \ln\frac{T_2}{T_1} - R\ln\frac{P_2}{P_1}$$

$$= 1.005 \ln \frac{340}{425} - 0.287 \ln\frac{1}{3} = \underline{0.091 \text{ kJ/kgK}}$$

Problem 1.9

Two alternative positions of the control surface may be used:

Case (i)

$T_w = 550$ K
$T_M = 590$ K

$$\dot{\Pi} = \dot{S}_{OUT} - \dot{S}_{IN} - \sum_i \frac{\dot{Q}_i}{T_i}$$

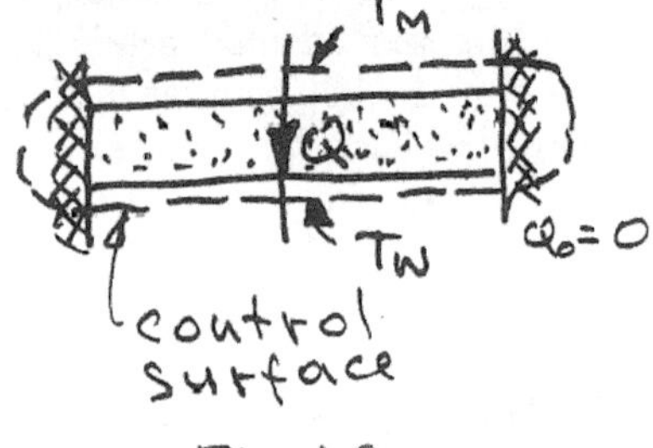

Fig. 1.9.a

with $\dot{S}_{OUT} = 0$ and $\dot{S}_{IN} = 0$

$$\dot{\Pi} = \frac{\dot{Q}}{T_W} - \frac{\dot{Q}}{T_M} = \dot{Q}\frac{T_M - T_W}{T_M T_W}$$

$$= 500 \text{ [MW]} \left[\frac{40}{550 \times 590}\right] \text{K}^{-1}$$

$$= \underline{61.6 \text{ kW/K}}$$

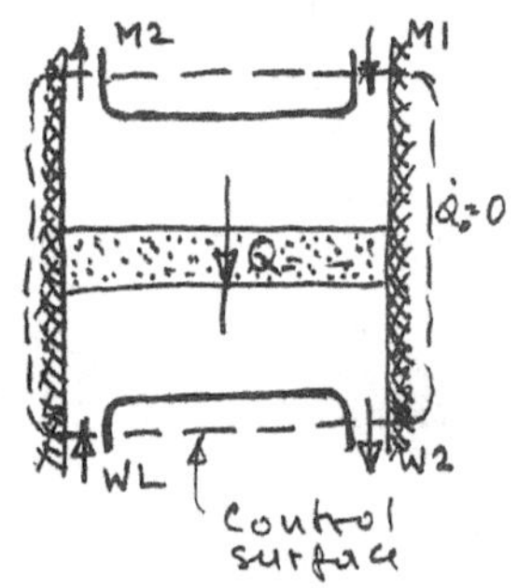

Fig. 1.9.b

Case (ii)

$$\dot{\Pi} = \dot{S}_{OUT} - \dot{S}_{IN} - \sum_i \frac{\dot{Q}_i}{T_i}$$

with $\dot{Q}_i = 0$

$$\dot{\Pi} = \left(\dot{S}_{M2} + \dot{S}_{W2}\right) - \left(\dot{S}_{M1} + \dot{S}_{W1}\right)$$

$$= \left(\dot{S}_{W2} + \dot{S}_{W1}\right) - \left(\dot{S}_{M1} + \dot{S}_{M2}\right)$$

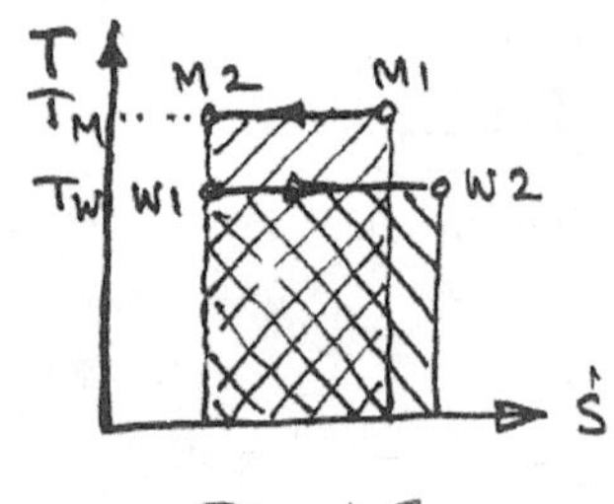

Fig. 1.9.c

But

$$\dot{Q} = T_M\left(\dot{S}_{M1} - \dot{S}_{M2}\right)$$

$$= T_W\left(\dot{S}_{W2} - \dot{S}_{W1}\right)$$

as shown on the T - S diagram

$$\therefore \dot{\Pi} = \frac{\dot{Q}}{T_W} - \frac{\dot{Q}}{T_M} = \underline{61.6 \text{ kW/K}}$$

Comment: Since there are no temperature or pressure gradients in the additional space enclosed by the Case (ii) control surface, there is no entropy produced there.

Problem 1.10

$P_1 = 8$ bar $\quad T_1 = 300$ °C
$P_2 = 2$ bar $\quad T_2 = 150$ °C

Entropy production rate in steady flow with $\dot{Q} = 0$

$$\dot{\Pi} = \dot{S}_2 - \dot{S}_1$$

or, per mass of steam

$$\pi = \dot{\Pi} / \dot{m} = s_2 - s_1$$

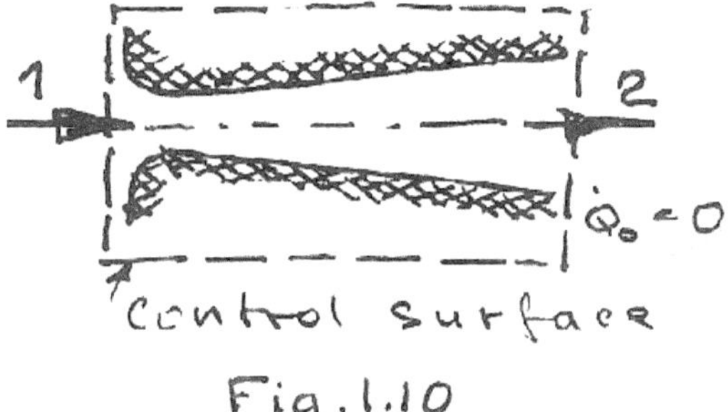

Fig.1.10

From t he steam tables

$s_1 = 7.233$ kJ/kgK
$s_2 = 7.280$ kJ/kgK

Substituting

$$\pi = \underline{7.280 - 7.233 = 0.047 \text{ kJ/kgK}}$$

Problem 1.11

(i) From the energy equation with

$$\dot{Q} = 0, \qquad \dot{W}_x = 0$$
$$\Delta KE = 0, \ \Delta PE = 0$$

$$\dot{m}_f(h_{fe} - h_{fi}) + m_s(h_{se} - h_{si}) = 0$$

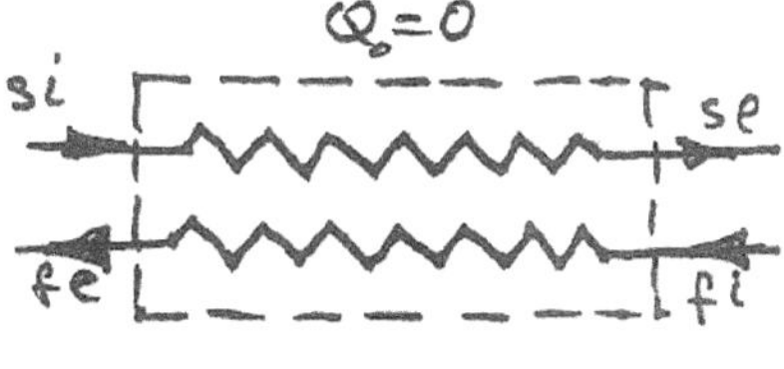

Fig.1.11.a

Given data:

		$\frac{P}{\text{bar}}$	T	$\frac{h}{\text{kJ/kg}}$	$\frac{s}{\text{kJ/kgK}}$
Heating Steam	At inlet	0.5	100 °C	2683	7.694
	At exit	0.5	(81.3 °C)	340	1.091
Feed Water	At inlet	10	75 °C	313.9	1.015
	At exit	10	95 °C	398	1.250

Hence,

$$\frac{\dot{m}_f}{\dot{m}_s} = \frac{h_{si} - h_{se}}{h_{fe} - h_{fi}} = \frac{2683 - 340}{298 - 313.9} = \underline{27.8}$$

(ii) Entropy production rate, with $\dot{Q}_o = 0$

$$\dot{\Pi} = \dot{S}_{OUT} - \dot{S}_{IN} = (\dot{m}_f s_{fe} + \dot{m}_s s_{se}) - (\dot{m}_f s_{fi} + \dot{m}_s s_{si})$$

Per mass of steam

$$\frac{\dot{\Pi}}{\dot{m}_s} = \frac{\dot{m}_f}{\dot{m}_s}(s_{fe} - s_{fi}) - (s_{si} - s_{se})$$

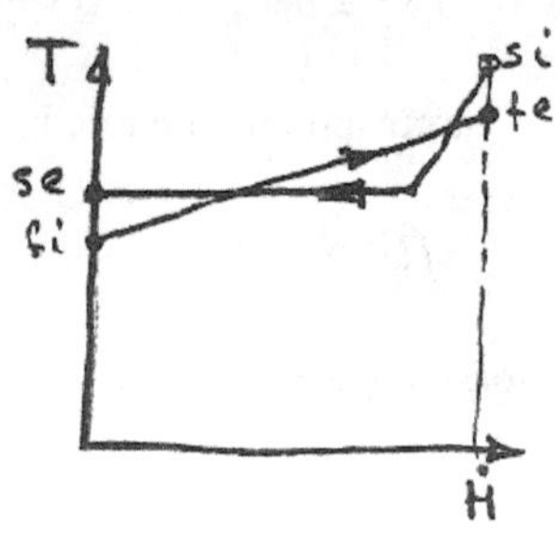

Fig.1.11.b

Substituting the given data,

$$\frac{\dot{\Pi}}{\dot{m}_s} = 27.8(1.250 - 1.015) - (7.694 - 1.091)$$

$$= \underline{-0.07\ \text{kJ/kg}}$$

Comments: Since entropy production must always have a positive value, the assumed operating conditions violate the Second Law. Note, however, that they do not violate the First Law. The reason for this can be seen on the $T - \dot{H}$ diagram.

Problem 1.12

The reaction is $CH_4 + 2\,O_2 \rightarrow CO_2 + 2H_2O$

Given data:

$$\left(\tilde{g}_f^o\right)_{H_2O,l} = -237180\ \text{kJ/kmol}$$

$$\left(\tilde{g}_f^o\right)_{CO_2,g} = -137150\ \text{kJ/kmol}$$

$$\left(\tilde{g}_f^o\right)_{CH_4,g} = -50790\ \text{kJ/kmol}$$

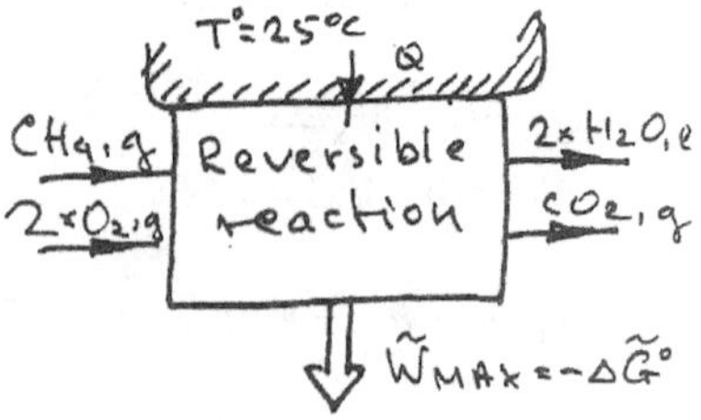

Fig. 1.12

Under conditions of reversibility

$\tilde{W}_{MAX} = -\Delta\tilde{G}$ where $\Delta\tilde{G}$, the Gibbs function of the reaction is expressible in terms of the Gibbs functions of formation, as follows:-

$$\Delta\tilde{G}^{o} = 2\left(\tilde{g}_{f}^{o}\right)_{H_2O,l} + \left(\tilde{g}_{f}^{o}\right)_{CO_2,g} - \left(\tilde{g}_{f}^{o}\right)_{CH_4,g}$$

Hence,

$$W_{MAX} = -(-50790 + 2\times 237180 + 137150)$$

$$= \underline{560720 \text{ kJ/kmol of } CH_4}$$

Problem 1.13

The reaction $H_2(g) + \frac{1}{2}O_2(g) \rightarrow H_2O(g)$

Maximum work $\qquad \tilde{W}_{MAX} = -\Delta\tilde{G}^{o}$

But $\quad \Delta\tilde{G}^{o} = -\ln K_p \times \tilde{R}T^{o}$

Also, $\quad \ln K_p = \dfrac{\log_{10} K_p}{0.4343}$

Given that

$$\log_{10} K_p = 40.048$$

$$\ln K_p = \frac{40.048}{0.4343} = 92.21$$

Hence, $\qquad \tilde{W}_{MAX} = \ln K_p \times \tilde{R}T^{o} = \underline{228595 \text{ kJ/mol}}$

Check $\qquad \tilde{W}_{MAX} = -\Delta\tilde{G}^{o} = -\left(\tilde{g}_{f}^{o}\right)_{H_2O,g} - \left(\tilde{g}_{f}^{o}\right)_{H_2,g} - \frac{1}{2}\left(\tilde{g}_{f}^{o}\right)_{O_2,g}$

$$= \underline{228590 \text{ kJ/kmol}}$$

Problem 2.1

Given data:

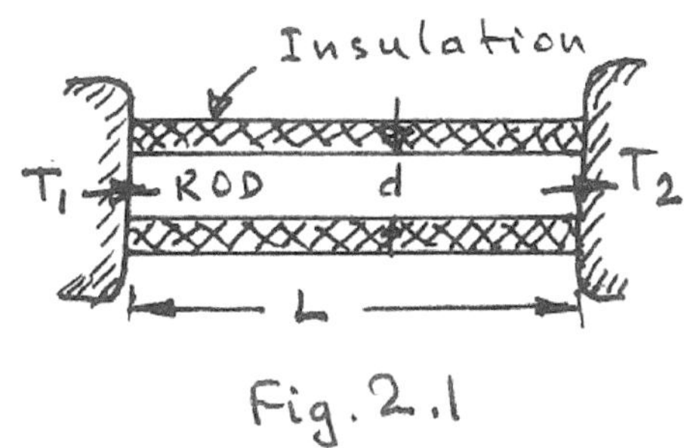

$T_1 = 100\ ^{o}C \qquad T_2 = 0\ ^{o}c$

$L = 0.5$ m $\qquad d = 0.02$ m

$K = 203$ W/(mK)

$T_o = 300$ K

Steady-state conduction along the rod

$$\dot{Q} = -kA\frac{T_2 T_1}{L}$$

$$= 203\left[\frac{\text{W}}{\text{mK}}\right]\frac{\pi 0.02^2}{4}[\text{m}^2]\frac{0-100}{0.5}\left[\frac{\text{K}}{\text{m}}\right]$$

$$= \underline{12.75\ \text{W}}$$

$\dot{E}^Q$ at the hot end

$$\dot{E}_1^Q = \dot{Q}\frac{T_1 - T_o}{T_1}$$

$$= 12.75\ \frac{373.15 - 300}{373.15}$$

$$= \underline{2.499\ \text{W}}$$

$\dot{E}^Q$ at the cold end

$$\dot{E}_2^Q = \dot{Q}\frac{T_2 - T_o}{T_2}$$

$$= 12.75\ \frac{273.15 - 300}{273.15}$$

$$= \underline{-1.253\ \text{W}}$$

Comment: Because the cold end is at a temperature lower than T_o, the exergy transfer takes place in a direction opposite to that of the heat transfer. Hence, the negative sign.

Problem 2.2

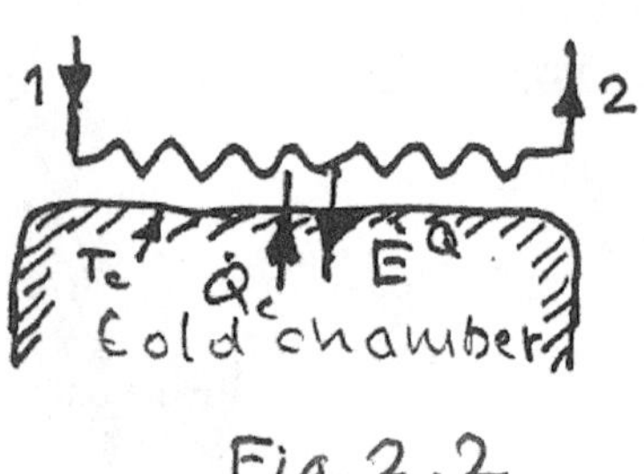

Fig. 2.2

Given data:

Refrigeration duty $\dot{Q} = 60$ kW

Cold chamber temperature $T_c = 250\ °\text{C}$

Environmental temperature $T_o = 300$ K.

Thermal exergy flow at the cold chamber temperature

$$\dot{E}_c^Q = \dot{Q}_c \frac{T_c - T_o}{T_c}$$

$$= 60\ [\text{kW}]\ \frac{250 - 300}{250} = \underline{-12\ \text{kW}}$$

Comment: The negative sign indicates that $\dot{E}_c^Q$ takes place in a direction opposite to that of $\dot{Q}_c$.

Problem 2.3

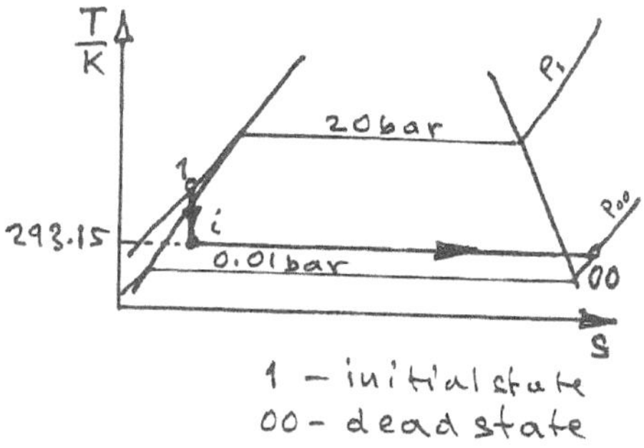

Given data:

$h_1 = 6.331$ kH/kg

$s_1 = 1.840$ kJ/kgK

$h_{oo} = 2535.2$ kJ/kg

$s_{oo} = 9.044$ kJ/kgK

$T_o = 293.15$ K

1 – initial state
00 – dead state

Fig. 2.3

$$\varepsilon_1 = (h_1 - h_{oo}) - T_o(s_1 - s_{oo})$$

$$= (6331.1 - 2535.2) - 293.15\ (1.840 - 9.044) = \underline{209.8 \frac{\text{kJ}}{\text{kg}}}$$

Processes:

1 - i - reversible adiabatic expansion
i - oo - reversible isothermal expansion at T_o.

where 1 - initial state
oo - dead state

Problem 2.4

In general, specific physical exergy is given by (note, for air $\varepsilon_o = 0$),

$$\varepsilon_{ph} = (h - h_o) - T_o(s - s_o)$$

In the case of a perfect gas,

$$\varepsilon_{ph} = c_P(T - T_o) - T_o(c_P \ln\frac{T}{T_o} - R\ln\frac{P}{P_o})$$

$$= \underbrace{c_P(T - T_o) - T_o \ln\frac{T}{T_o}}_{\varepsilon\Delta T} + \underbrace{RT_o \ln\frac{P}{P_o}}_{\varepsilon\Delta P}$$

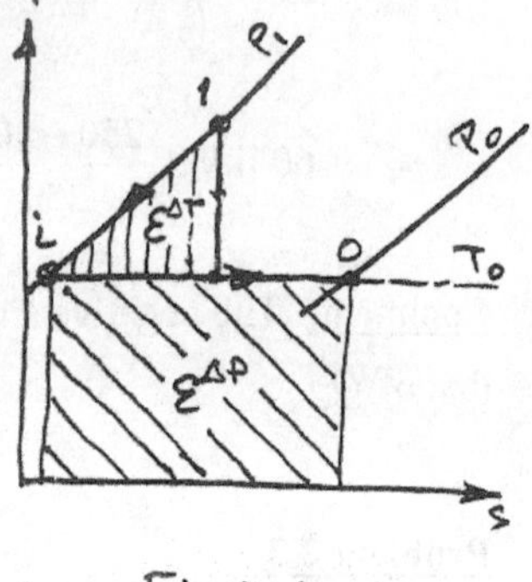

Fig.2.4.a

Case (i)

$P_1 > P_o$, $\quad T_1 > T_o$

$P_1 = 3$ bar $\quad T_1 = 400$ K

$T_o = 290$ K $\quad P_o = 1$ bar

Substituting

$$\varepsilon_{ph,1} = 1.005\left[400 - 290 - 290\ln\frac{400}{290}\right] + 290 \times 0.287\ln 3$$

$$= 16.82 + 91.44$$

$$= \underline{108.26 \text{ kJ/kg}}$$

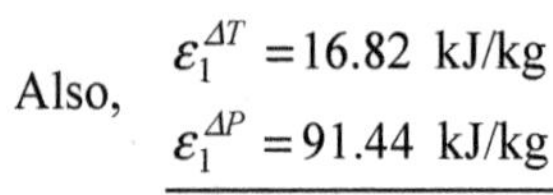

Also, $\varepsilon_1^{\Delta T} = 16.82$ kJ/kg
$\varepsilon_1^{\Delta P} = 91.44$ kJ/kg

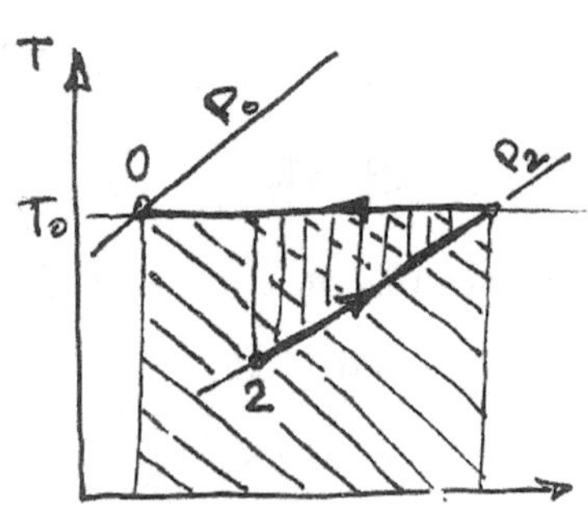

Fig.2.4.b

Case (ii)

$P_2 < P_o$, $\quad T_2 < T_o$

$P_2 = 0.8$ bar $\quad T_2 = 270$ K

Substituting

$$\varepsilon_{ph,2} = 1.005\left[270 - 290 - 290\ln\frac{270}{290}\right] + 290 \times 0.287\ln 0.8$$

$$= 0.72 - 18.58$$

$$= -17.86 \text{ kJ/kg}$$

Also, $\varepsilon_2^{\Delta T} = 0.72$ kJ/kg
$\varepsilon_2^{\Delta P} = -18.58$ kJ/kg

Problem 2.5

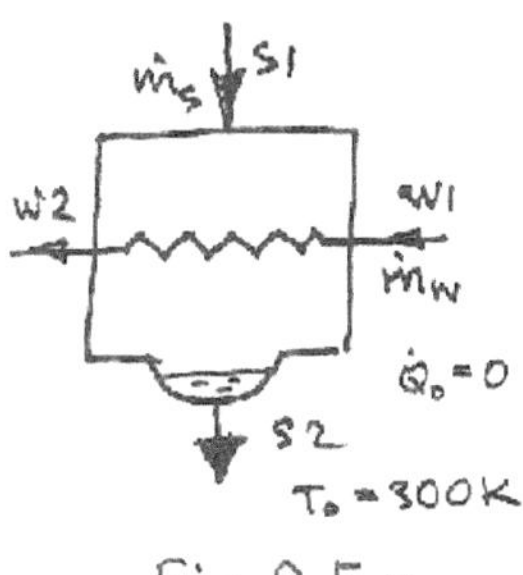

Fig. 2.5.a

Given data:

$\dot{m}_s = 0.1$ kg/s $\qquad P_s = 10$ bar

$T_{w1} = 20$ °C $\qquad T_{w2} = 165$ °C

Properties:

State	h kJ/kg	s kJ/kg K
s1	2778	6.586
s2	763	2.138
w1	83.9	0.296
w2	696	1.992

Mass flow rate, $\dot{m}_w$, from the energy balance with

$\dot{Q}_o = 0, \quad \dot{W}_x = 0, \quad \Delta KE = 0, \quad \Delta PE = 0$

$$\dot{m}_s = (h_{s1} - h_{s2}) = \dot{m}_w (h_{w2} - h_{w1})$$

$$\therefore \ \dot{m}_w = \frac{0.1(2778 - 763)}{696 - 83.9} = 0.329 \text{ kg/s}$$

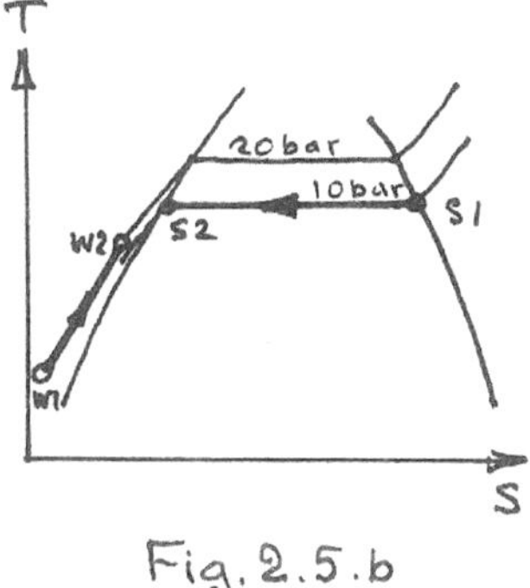

Fig. 2.5.b

Exergy transfer rate for a stream

$$\Delta \dot{E} = \dot{m}[(h_{OUT} - h_{IN}) - T_o(s_{OUT} - s_{IN})]$$

Steam

$$\Delta \dot{E}_s = 0.1[(763 - 2778) - 300(2.138 - 6.586)]$$

= <u>-68.08 KW</u> i.e. suffers a reduction in exergy flow rate.

Water

$$\Delta \dot{E}_w = 0.329[(696 - 83.9) - 300(1.992 - 0.296)]$$

= <u>34.31 kW</u> i.e. undergoes an increase in its exergy flow rate.

Problem 2.6

Given data:

$\dot{m} = 0.5$ kg/s $P = 50$ bar = const.

$T_1 = 30$ °C $T_2 = 400$ °C $T_o = 300$ K

Properties:

State	h kJ/kg	s kJ/kg K
1	125.7	0.436
2	3196	6.646

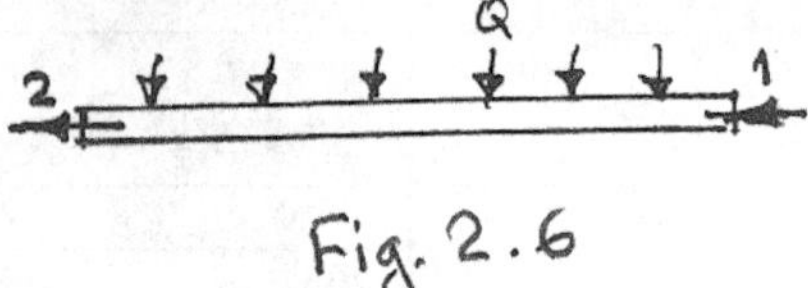

Fig. 2.6

Change in exergy flow rate:

$$\dot{E}_2 - \dot{E}_1 = \dot{m}[(h_2 - h_1) - T_o(s_2 - s_1)]$$

$$= 0.5\ [(3196 - 125.7) - 300\ (6.646 - 0.436)]$$

$$= \underline{603.65\ \text{kW}}$$

Problem 2.7

Standard chemical exergy of a reference substance = reversible isothermal work at T^o

$$\therefore \tilde{\varepsilon}_i^o = \tilde{R}T^o \ln \frac{P^o}{P_i^{oo}} \qquad T^o = 298.15\ \text{K}$$

In the case of an ideal gas $P_i^{oo} / P^o = x_i^{oo}$

Hence, $\tilde{\varepsilon}_i^o = \tilde{R}T^o \ln(x_i^{oo})^{-1}$

x_i^{oo} - standard mole fraction of the gas in the atmosphere

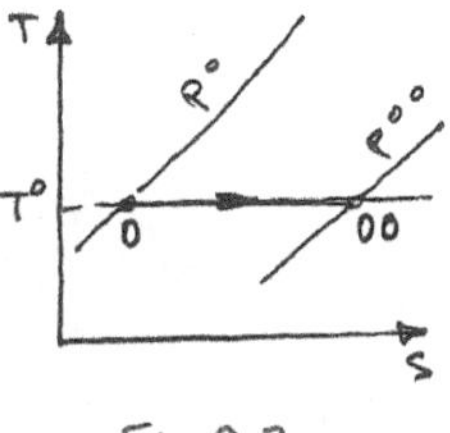

Fig. 2.7

Given data:

	O_2	CO_2	H_2O_{vap}
$x_i^{oo} \times 10^3$	20.40	0.294	8.8
x_i	0.52	0.45	0.03

Standard molar chemical exergies

$\underline{O_2}$ $\tilde{\varepsilon}_{O2}^o = 8.3143 \times 298.15 \ln (0.0204)^{-1}$

= 3940 kJ/kmol

$\underline{CO_2}$ $\quad \tilde{\varepsilon}^{o}_{CO_2} = 8.3143 \times 298.15 \ln (0.000294)^{-1}$

$= \underline{20158 \text{ kJ/kmol}}$

$\underline{H_2O}$ $\quad \tilde{\varepsilon}^{o}_{H_2O} = 8.3143 \times 298.15 \ln (0.0088)^{-1}$

$= \underline{11733 \text{ kJ/kmol}}$

Standard molar chemical exergy of the specified mixture

$$\tilde{\varepsilon}^{o}_{M} = \sum_{i} x_i \tilde{\varepsilon}^{o}_{i} + \tilde{R}T^{o} \sum_{i} x_i \ln x_i$$

Substituting the given numerical values

$\tilde{\varepsilon}^{o}_{M} = \underline{9477 \text{ lK/kmol}}$

Problem 2.8

Standard molar chemical exergy of a chemical substance (see p.238, eq. A.7)

$$\tilde{\varepsilon}^{o} = -\Delta G^{o} - \sum_{j} n_j \tilde{\varepsilon}^{o}_{j} + \sum_{k} n_k \tilde{\varepsilon}^{o}_{k}$$

Subscripts j and k are for co-reactants and products of the reference reaction, respectively.

Reference reaction for CH_4

$$CH_4 + 2O_2 \rightarrow CO_2 + 2H_2O$$

Given data:

	CH_4	CO_2	H_2O_{vap}
$\Delta \tilde{g}^{o}_{f}$ /[kJ/kmol]	-50810	-394390	-228590

Hence, the Gibbs function of the reference reaction

$$\left(\Delta G^{o}\right)_{CH_4} = \left(\Delta \tilde{g}^{o}_{f}\right)_{CO_2} + 2\left(\Delta \tilde{g}^{o}_{f}\right)_{H_2O} - \left(\Delta \tilde{g}^{o}_{f}\right)_{CH_4}$$

$$= -394390 - 2 \times 228590 + 50810$$
$$= -800760 \text{ kJ/kmol}$$

$\tilde{\varepsilon}^{o}$ for CH_4 (using $\tilde{\varepsilon}^{o}_{CO_2}$, $\tilde{\varepsilon}^{o}_{H_2O}$, $\tilde{\varepsilon}^{o}_{O_2}$ from Problem 2.7)

$$(\tilde{\varepsilon}^{o})_{CH_4} = -(\Delta G^{o})_{CH_4} + \tilde{\varepsilon}^{o}_{CO_2} + 2\times\tilde{\varepsilon}^{o}_{H_2O} - 2\times\tilde{\varepsilon}^{o}_{o2}$$

$$= 800760 + 20158 + 2 \times 11733 - 2 \times 3940$$

$$= \underline{836504 \text{ kJ/kmol}}$$

Problem 2.9

Molar chemical exergy of a mixture per mole of the mixture is given by:-

$$\tilde{\varepsilon}_{OM} = \sum_{i} x_i \tilde{\varepsilon}_{oi} + \tilde{R}T_o \sum_{i} x_i \ln x_i$$

Given data:

$T_o = 25$ °C

	x_i	$\tilde{\varepsilon}_{oi}$ /[kJ/kmol]
CH_4	1/21	836504
Air	20/21	0

Substituting

$$\tilde{\varepsilon}_{OM} = \frac{20}{21}\times 0 + \frac{1}{21}\times 836504$$

$$+ 8.3143\times 298.15\left[\frac{20}{21}\ln\frac{20}{21} + \frac{1}{21}\ln\frac{1}{21}\right]$$

$$= \underline{39358.9 \text{ kJ/kmol}}$$

Problem 2.10

Given data:

$$\Delta\tilde{g}^{o}_{f,CO} = -137160 \text{ kJ/kmol}$$

$$\tilde{\varepsilon}^{o}_{O_2} = 3975 \text{ kJ/kmol}$$

$$\tilde{\varepsilon}^{o}_{c} = 410530 \text{ kJ/kmol}$$

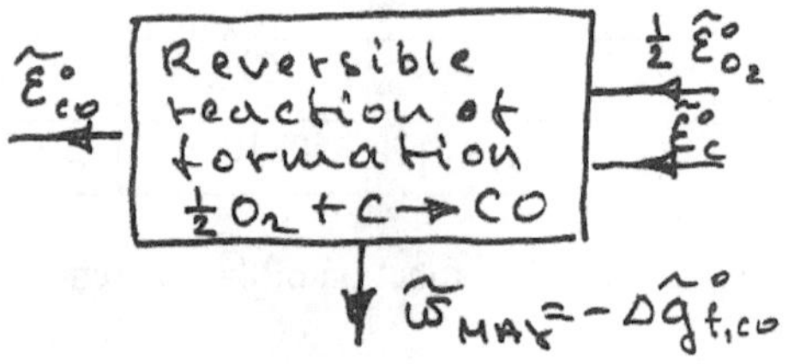

Fig. 2.10

To calculate $\tilde{\varepsilon}^{o}_{c}$

Since there is no degradation of energy in a reversible reaction

$$\tilde{\varepsilon}_{\mathrm{c}}^{\mathrm{o}} = \Delta\tilde{g}_{\mathrm{f,CO}}^{\mathrm{o}} + \frac{1}{2}\tilde{\varepsilon}_{\mathrm{O2}}^{\mathrm{o}} + \tilde{\varepsilon}_{\mathrm{c}}^{\mathrm{o}}$$

$$= -137160 + \frac{1}{2} \times 3975 + 410530$$

$$= \underline{275357.5 \text{ kJ/kmol}}$$

Problem 2.11

Standard molar chemical exergy of a mixture of gases

$$\tilde{\varepsilon}_{\mathrm{M}}^{\mathrm{o}} = \sum_i x_{\mathrm{i}}\tilde{\varepsilon}_{\mathrm{i}}^{\mathrm{o}} + \tilde{R}T^{\circ}\sum_i x_{\mathrm{i}} \ln x_{\mathrm{i}}$$

Given data:

$T_{\mathrm{o}} = 298.15$ K

	$x_{\mathrm{i}} \times 100$	$\tilde{\varepsilon}_i^o$ /[kJ/kmol]
CO	4.09	275350
N_2	69.30	720
H_2O	20.48	11770
CO_2	6.13	20170

Substituting in the expression

$$\tilde{\varepsilon}_{\mathrm{M}}^{\mathrm{o}} = \underline{13224.5 \text{ kJ/kmol}}$$

Problem 2.12

Standard molar chemical exergy of a mixture of gases

$$\tilde{\varepsilon}_{\mathrm{M}}^{\mathrm{o}} = \sum_i x_{\mathrm{i}}\tilde{\varepsilon}_{\mathrm{i}}^{\mathrm{o}} + \tilde{R}T^{\circ}\sum_i x_{\mathrm{i}} \ln x_{\mathrm{i}}$$

Given data:

	$x_i \times 100$	$\tilde{\varepsilon}_i^o$ /[kJ/kmol]
H_2	47.0	238350
CH_4	41.0	836510
C_2H_4	2.5	1366610
CO	7.5	275350
N_2	2.0	720

Substituting in the expression

$$\tilde{\varepsilon}_M^o = \underline{507134.5 \text{ kJ/kmol}}$$

Problem 2.13

Exergy of the mixture = reversible work obtained from all the N components, i.e.

$$\dot{E}_1 = \sum_N \left[(\dot{W}_x)_{REV}\right]_i$$

$$= \sum \dot{n}_i \left[(\tilde{h}_{i1} - \tilde{h}_{ioo}) - T_o(\tilde{s}_{i1} - \tilde{s}_{ioo})\right]$$

For perfect gases

$$\tilde{h}_{i1} - \tilde{h}_{ioo} = \tilde{c}_{Pi}(T_1 - T_o)$$

$$\tilde{s}_{i1} - \tilde{s}_{ioo} = \tilde{c}_{Pi} \ln\frac{T_1}{T_o} - \tilde{R}\ln\frac{P_{i1}}{P_{ioo}}$$

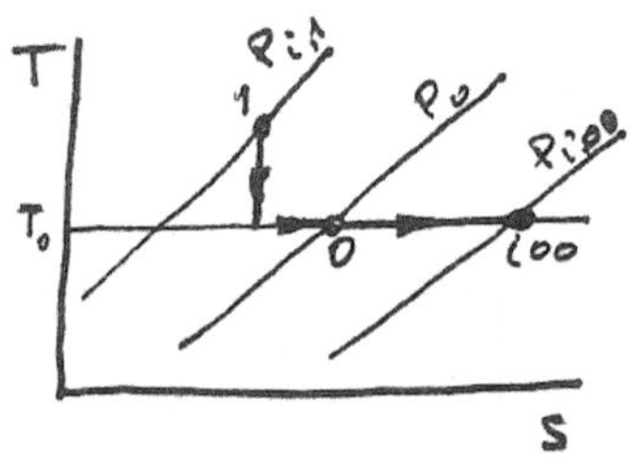

Fig.2.13.a

Also, $P_{i1} = P_1 x_i$, $\quad P_{ioo} = P_o x_{ioo}$, $\quad x_i = \dfrac{\dot{h}_{i1}}{\dot{h}_1}$

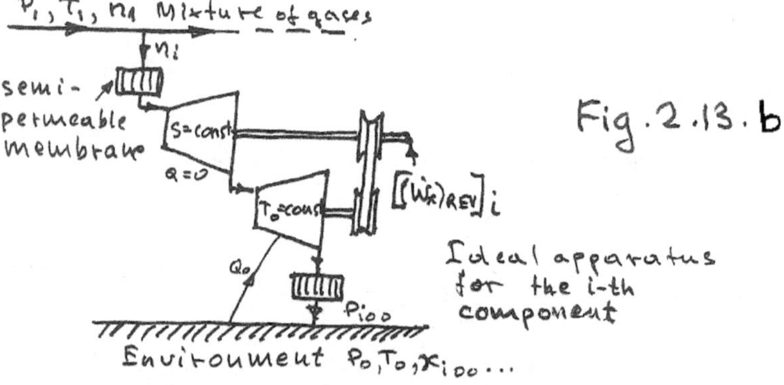

Fig.2.13.b

Hence,

$$\tilde{\varepsilon}_1 = \frac{\dot{E}_1}{\dot{h}_1} = \sum_N x_i \left\{ \tilde{c}_{Pi}\left[(T_1 - T_o) - T_o \ln\frac{T_1}{T_o}\right] + \tilde{R}T_o \ln\frac{P_1 x_i}{P_o x_{ioo}} \right\}$$

Numerical calculations

Since $P_1 = P_o$ the above equation can be written as follows:

$$\dot{E}_1 = \dot{h}\tilde{\varepsilon}_1 = \dot{h}[f(T)]\sum_N (x_i \left\{\tilde{c}_{Pi}) + \tilde{R}T_o \sum x_i \ln \frac{x_i}{x_{ioo}}\right\}$$

where $f(T) = T_1 - T_o - T_o \ln(T_1 / T_o)$

$$= 250 - 25 - 298.15 \ln \frac{523.15}{298.15} = 57.36 \text{ K}$$

From given data:

Constit. of mixture	x_i	x_{ioo}	$\frac{\tilde{c}_{Pi}}{\text{kJ/kmol K}}$	$\frac{x_i \tilde{c}_{Pi}}{\text{kJ/kmol K}}$	$x_i \ln \frac{x_i}{x_{ioo}}$
CO	0.05	0.0003	40.0	2.000	0.2560
H_2O	0.12	0.0177	34.0	4.080	0.2300
O_2	0.08	0.2060	29.9	2.392	-0.0757
N_2	0.75	0.7760	29.2	21.900	-0.0256
Total				30.372	0.3847

Substituting

$$\dot{E}_1 = 0.12\left[\frac{\text{kmol}}{\text{s}}\right][57.36[\text{K}]30.372\left[\frac{\text{kJ}}{\text{kmol K}}\right]$$

$$+ 8.3143\left[\frac{\text{kJ}}{\text{kmol K}}\right]0.3847 \times 298.15[K]]$$

$= \underline{323.5 \text{ kW}}$

Problem 2.14

Given data:

A stream of a combustible gas mixture specified in Problem 2.9, plus

$\dot{m}$ = 0.5 kg/s, T_1 = 100 °C P_1 = 1.2 bar

C_1 = 100 m/s, P_o = 1 bar T_o = 25 °C

	CH_4	Air
Molar mass $\tilde{m}$ / [kg/kmol]	16.042	28.96
Sp. Heat capacity c_p / [kJ/kg K]	2.334	1.006

From Problem 2.9 $\tilde{\varepsilon}_{OM} = 39358.9$ kJ/kmol

Exergy flow rate of a stream

$$\dot{E}_1 = \dot{m}\left[\varepsilon_{\mathrm{k},1} + \varepsilon_{\mathrm{ph},1} + \varepsilon_{\mathrm{o}}\right]$$

Kinetic exergy (specific)

$$\varepsilon_{\mathrm{k},1} = \frac{C_1^2}{2} = \frac{100^2}{2} \times 10^{-3} = 5 \text{ kJ/kg}$$

Physical exergy (specific), perfect gas

$$\varepsilon_{\mathrm{ph},1} = c_P (T_1 - T_{\mathrm{o}}) - T_{\mathrm{o}}\left(c_P \ln \frac{T_1}{T_{\mathrm{o}}} - R \ln \frac{P_1}{P_{\mathrm{o}}} \right)$$

$$c_P = \frac{\widetilde{c}_P}{\widetilde{m}} \qquad R = \frac{\widetilde{R}}{\widetilde{m}}$$

$$\widetilde{m} = \sum x_{\mathrm{i}} \widetilde{m}_{\mathrm{i}} = \frac{20}{21} 28.96 + \frac{1}{21} 16.042$$

$$= 28.34 \text{ kg/kmol}$$

$$\widetilde{c}_P = \sum x_i \widetilde{c}_{P_i} = \frac{20}{21} \times 1.006 \times 28.96 + \frac{1}{21} 2.334 \times 16.042$$

$$= 29.53 \text{ kJ/kmol}$$

$$\therefore c_P = \frac{29.53}{28.34} = 1.042 \text{ kJ/kg K}$$

$$R = \frac{8.3144}{28.34} = 0.2934 \text{ kJ/kg K}$$

Hence, physical exergy is calculated

$$\varepsilon_{\mathrm{ph},1} = 1.042(100 - 25) - 298.15\left(1.042 \ln \frac{373.15}{298.15} - 0.2934 \ln \frac{1.2}{1.0} \right)$$

$$= 13.1 \text{ kJ/kg}$$

Chemical exergy (specific)

$$\varepsilon_{\mathrm{OM}} = \frac{\widetilde{\varepsilon}_{\mathrm{OM}}}{\widetilde{\mathrm{m}}} = \frac{39358.9}{28.34} = 1388.8 \text{ kJ/kg}$$

Exergy flow rate of the stream $\dot{E}_1 = 0.5[5 + 13.1 + 1388.8]$

$= \underline{703.46 \text{ kW}}$

Problem 2.15

Since air is a common substance in thermodynamic, unrestricted equilibrium with the environment, its chemical non-flow exergy is zero, $\xi_o = 0$, and therefore $\xi = \xi_{ph}$.
In general,

$$\xi_{ph,1} = (U_1 - U_o) + P_o(v_1 - v_o) - T_o(s_1 - s_o)$$

For air, when treated as a perfect gas

$$\xi_1 = \xi_{ph,1} = c_v(T_1 - T_o) + R\left(T_1 \frac{P_o}{P_1} - T_o\right) - T_o(s_1 - s_2)$$

$$= RT_o\left[\frac{1}{\gamma-1}\left(\frac{T_1}{T_o} - 1\right) + \left(\frac{T_1}{T_o}\frac{P_o}{P_1} - 1\right) - \left(\frac{1}{\gamma-1}\ln\frac{T_1}{T_o} - \ln\frac{P_1}{P_o}\right)\right] \qquad \text{(a)}$$

Case (i) $\quad P_1 = 3$ bar $\quad T_1 = 400$ K

$\quad P_o = 1$ bar $\quad T_o = 290$ K

Substituting

$$\xi_1 = 0.2821 \times 290\left[\frac{1}{0.4}\left(\frac{400}{290} - 1\right) + \left(\frac{400}{290}\frac{1}{3} - 1\right) - \left(3.5\ln\frac{400}{290} - \ln 3\right)\right]$$

$= \underline{31.7 \text{ kJ/kg}}$

Alternatively, (see p.52)

$$\xi_1 = \varepsilon_1 - (P - P_o)v_1$$

for a perfect gas,

$$\xi_1 = \varepsilon_1 - RT_1\left(1 - \frac{P_o}{P_1}\right) \qquad \text{(b)}$$

Using $\varepsilon_1 =$ 108.29 kJ/kg from Problem 2.4

$$\xi_1 = 108.29 - 0.2871 \times 400\left(1 - \frac{1}{3}\right)$$

$= \underline{31.65 \text{ kJ/kg}}$ (as above)

Case (ii) $\quad P_2 = 0.8$ bar $\quad T_2 = 270$ K

$\quad P_o = 1$ bar $\quad T_o = 290$ K

Substituting in eq. (a)

$$\underline{\xi_2 = 1.52 \text{ kJ/kg}}$$

An identical answer is obtained when using eq. (b), as is to be expected.

Note, that in both cases, the values of ξ are positive, unlike the values of ε calculated in Problem 2.4. This confirms that non-flow exergy has always positive value.

Problem 2.16

Exergy of a closed system (i.e. non-flow exergy) is given by

$$\Xi_1 = m[(u_1 - u_o) + P_o(v_1 - v_o) - T_o(s_1 - s_o)]$$

for a perfect gas, this becomes

$$\Xi_1 = P_o V_1 \frac{T_o}{T_1}\left[\frac{1}{\gamma - 1}\left(\frac{T_1}{T_o} - 1\right)\frac{P_1}{P_o} + \left(\frac{T_1}{T_o} - \frac{P_1}{P_o}\right) - \frac{P_1}{P_o}\left(\frac{\gamma}{\gamma - 1}\ln\frac{T_1}{T_o} - \ln\frac{P_1}{P_o}\right)\right] \quad \text{(a)}$$

Given data:

Pressure vessel capacity $V_1 = 0.1 \text{ m}^3$

$P_1 = 5$ bar $T_1 = 400$ K $\gamma = 1.4$

$P_o = 1$ bar $T_o = 300$ K

Substituting in eq. (a)

$$\Xi_1 = 10^5 \frac{N}{\text{m}^2} 0.1\,\text{m}^3 \frac{300}{400}\left[\frac{1}{1.4 - 1}\left(\frac{400}{300} - 1\right)5 + \left(\frac{400}{300} - 5\right) - 5\left(\frac{1}{1.4 - 1}\ln\frac{400}{300} - \ln 5\right)\right]$$

$$= \underline{26.32 \text{ kJ}}$$

From eq. (a), when $T_1 \to T_o$ and $P_1 \to 0$

$$\Xi_1 = P_o V_1$$

$$= 10^5 \frac{N}{\text{m}^2} 0.1\,\text{m}^3 = \underline{10 \text{ KJ}}$$

Comment: This demonstrates the fact that exergy can be regarded as a measure of disequilibrium between the system and the environment.

Problem 2.17

Derivation of an expression for the exergy of a stream of humid air.

As shown in Problem 2.13, exergy of a mixture made up from environmental substances is given by

$$\dot{E}_1 = \sum_N \dot{n}_{i1}\left[\left(\tilde{h}_{i1} - \tilde{h}_{ioo}\right) - T_o\left(\tilde{s}_{i1} - \tilde{s}_{ioo}\right)\right]$$

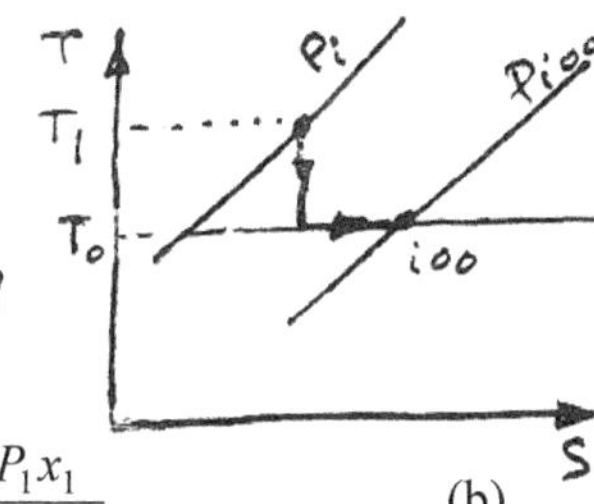

Fig. 2.17

For perfect gases this becomes

$$\tilde{\varepsilon}_1 = \frac{\dot{E}_1}{\dot{n}_1} = \sum_N x_i\left\{\tilde{c}_{Pi}\left(T_1 - T_o\right) - T_o \ln\frac{T_1}{T_o}\right\} + \tilde{R}T_o \frac{P_1 x_1}{P_o x_{ioo}} \qquad \text{(b)}$$

For the case of a two component mixture, air and water vapour, x_a, x_v,

$$\tilde{\varepsilon}_1 = \left(x_a \tilde{c}_{P,a} + x_v \tilde{c}_{p,v}\right)T_o\left[\frac{T_1}{T_o} - 1 - \ln\frac{T_1}{T_o}\right] + \tilde{R}T_o \ln\frac{P_1}{P_o}$$

$$+ \tilde{R}T_o\left(x_a \ln\frac{x_a}{x_{aoo}} + x_v \ln\frac{x_v}{x_{voo}}\right) \qquad \text{(c)}$$

We, further, have

$$x_a + x_v = 1 \quad \text{and} \quad x_{aoo} + x_{voo} = 1 \qquad \text{(d)}$$

mole fraction ratio $\quad \tilde{\omega}_1 = \dfrac{x_v}{x_a} \qquad$ (e)

and for the atmosphere $\quad \tilde{\omega}_{oo} = \dfrac{x_{voo}}{x_{aoo}} \qquad$ (f)

From (d), (e) and (f)

$$x_a = \frac{1}{1+\tilde{\omega}_1}, \quad x_v = \frac{\tilde{\omega}_1}{1+\tilde{\omega}_1}, \quad x_{aoo} = \frac{1}{1+\tilde{\omega}_{oo}}, \quad x_v = \frac{\tilde{\omega}_{oo}}{1+\tilde{\omega}_{oo}}$$

Substituting these in (c), we get

$$\tilde{\varepsilon}_1 = \frac{\tilde{c}_{P,a} + \tilde{\omega}_1 c_{P,v}}{1+\tilde{\omega}_1} T_o\left(\frac{T_1}{T_o} - 1 - \ln\frac{T_1}{T_o}\right) + \tilde{R}T_o \ln\frac{P_1}{P_o}$$

$$+ \tilde{R}T_o\left[\ln\frac{1+\tilde{\omega}_{oo}}{1+\tilde{\omega}_1} + \frac{\tilde{\omega}_1}{1+\tilde{\omega}_1}\ln\frac{\tilde{\omega}_1}{\tilde{\omega}_{oo}}\right]$$

as required.

Problem 3.1

The process: An initially evacuated vessel is filled with air from the atmosphere.

To determine: The irreversibility of the process.

Given data: $V_1 = 0.1\ \text{m}^3 \quad P_o = 1$ bar

Assumptions: the vessel is rigid and adiabatic.

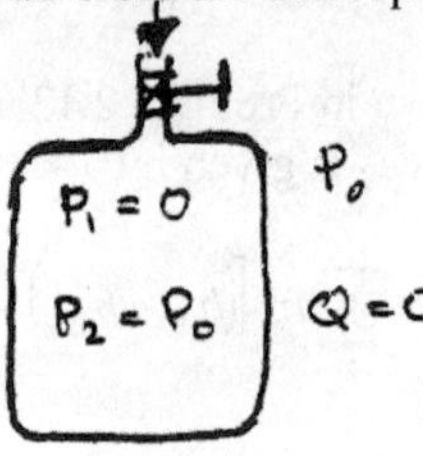

Analysis: The exergy balance for a closed system

$$\Xi_1 = \Xi^Q = \Xi_2 + W_{NET} + I \qquad \text{(a)}$$

Fig. 3.1

where

$$\Xi = \Xi_{ph} + \Xi_o \quad \text{for air } \Xi_o = 0 \qquad \text{(b)}$$

$$\therefore \quad \Xi_{ph} = [(u + P_o v - T_o s) - (u_o + P_o v_o - T_o s_o)]\text{m} \qquad \text{(c)}$$

$$\text{or} \quad \Xi_{ph} = \text{m}(u\text{-}u_o) + P_o V - P_o v_o \text{m} = \text{m}T_o(s - s_o) \qquad \text{(d)}$$

$$\Xi^Q = \sum_r \left[Q \frac{T_r - T_o}{T_r} \right] \quad \text{but since } Q = 0 \ \Xi^Q = 0$$

Since for state 2, $P_2 = P_6$, $u_2 = u_o$ $v_2 = v_o$ we have $\Xi_2 = 0$.

Also, $W_{NET} = 0$

Hence, $I = \Xi_1$

But, since in state 1, m = 0, it follows from (d) that

$$I = P_o V_1 = 10^5 \frac{N}{\text{m}^2} 0.1\,\text{m}^3 = 10^4 \text{ Nm}$$

$$= \underline{10 \text{ kJ}}$$

Compare this answer with that of the second part of Problem 2.16.

Problem 3.2

The process: Escape of CO_2 from a vessel into the atmosphere.

To determine: The destruction of exergy (or irreversibility) caused by the escape of CO_2.

Given data: $V = 0.1\ \text{m}^3$ $\quad P_1 = 30$ bar $\quad T_1 = 10\ °\text{C}$

$T_o = 10\ °\text{C}$ $\quad P_o = 1$ bar $\quad (x_{oo})_{CO_2} = 0.0003$

Assumptions: CO_2 behaves as an ideal gas.

Fig.3.2.a

Analysis: All the initial exergy of CO_2 is destroyed as a result of the escape.

Therefore $\quad I = \Xi_1$ (a)

But (see book p.52)

$$\Xi_1 = E_1 - (P_1 - P_o)V_1 \tag{b}$$

and

$$E_1 = E_{\text{ph},1} + E_{\text{o}.1} \tag{c}$$

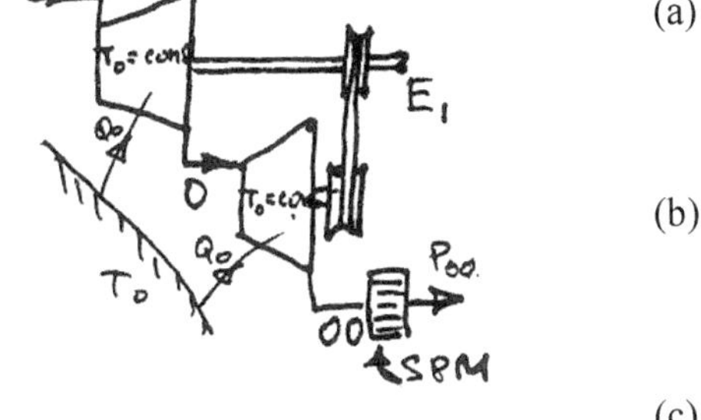

Fig.3.2.b

The flow exergy, E_1, is equal to reversible isothermal work obtained from the steady flow expansion processes which are illustrated above, i.e.

$$E_1 = n\tilde{R}T_o\left(\ln\frac{P_1}{P_o} + \ln\frac{P_o}{P_{oo}}\right) = n\tilde{R}T_o \ln(P_1 / P_{oo} \tag{d}$$

But $\quad n\tilde{R}T_o = P_1V_1$ (e)

Hence, from the above

$$I = P_1V_1 \ln\frac{P_1}{P_{oo}} - (P_1 - P_o)V_1$$

$$= P_1V_1[\ln(P_1 - P_{oo}) - 1 + (P_o - P_1)]$$

$$= 30\times10^5\ \frac{N}{\text{m}^2}\times 0.1\ \text{m}^3\left[\ln\frac{30}{0.0003} - 1 + \frac{1}{30}\right] = \underline{3163.9\ \text{kJ}}$$

Problem 3.3

The process: Frictionless polytropic compression in a piston and cylinder assembly.

To determine: The irreversibility of the process.

Given data: m = 0.0025 kg

$P_1 = 1$ bar $\quad T_1 = 20$ °C

$P_2 = 4$ bar $\quad T_o = 293.15$ K $\quad PV^{1.25} = \text{const.}$

$$T_2 = T_1\left(\frac{P_2}{P_1}\right)^{\frac{0.25}{1.25}} = 386.8\text{ K}$$

Fig.3.3.a

Assumptions: air behaves as a perfect gas.

Analysis: We use the Gouy-Stodola relation

$$I = T_o\left[(S_2 - S_1) - \sum_i \frac{Q_i}{T_i}\right] \tag{a}$$

But here $Q_i = -Q_o$ and $T_i = T_o$

$$\therefore I = T_o \text{m}(s_2 - s_1) + Q_o \tag{b}$$

The energy balance for a polytropic process

$$Q = W + U_2 - U_1$$
$$= \text{m}\left[\frac{R}{1-n}(T_2 - T_1) + c_v(T_2 - T_1)\right]$$
$$= \text{m}\, c_v \frac{\gamma - n}{1-n}(T_2 - T_1) \tag{c}$$

Also, $$s_2 - s_1 = c_P \ln\frac{T_2}{T_1} - R\ln\frac{P_2}{P_1} \tag{d}$$

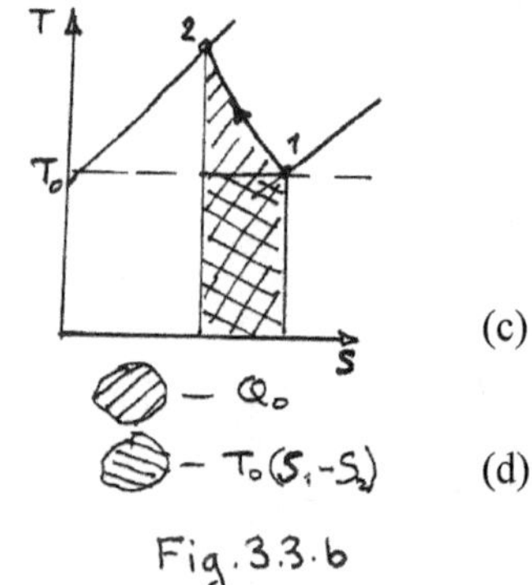

Fig.3.3.b

Substituting (c) and (d) in (b)

$$I = \text{m}\, c_v\left\{T_o\left[\gamma \ln\frac{T_2}{T_1} - (\gamma - 1)\ln\frac{P_2}{P_1}\right] + \frac{\gamma - n}{n-1}(T_2 - T_1)\right\}$$

$$= 0.0025\text{ kg} \times 0.718\frac{\text{kJ}}{\text{kg K}}\{293.15\text{ K}\left[1.4\ln\frac{386.8}{293.15} - (1.4-1)\ln 4\right]$$

$$+ \frac{1.4 - 1.25}{1.25 - 1}(386.8 - 2.93.15)\text{ K}\}$$

$$= \underline{0.0133\text{ kJ}}$$

Comments: Note that the process is frictionless and therefore internally reversible.

The irreversibility calculated is of external type and is due to heat transfer over a finite temperature difference between the air inside the cylinder at temperature $T > T_o$ and the environment at T_o. As follows from eq. (b), the irreversibility can be represented on the T-S diagram by the difference in the two cross-hatched areas.

Problem 3.4

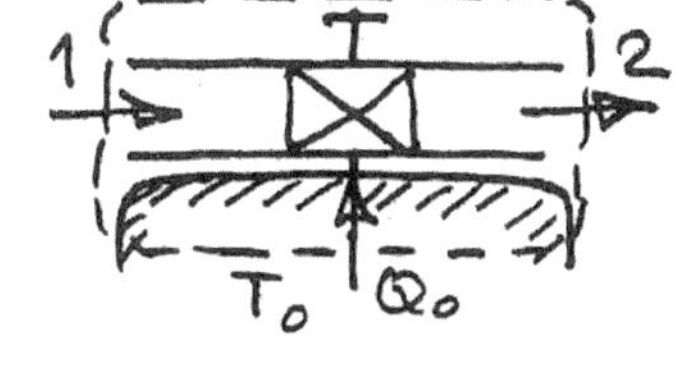

Fig.3.4.a

The process: Throttling process at $T < T_o$

To determine: (i) Heat transfer, $\dot{Q}_o$,

(ii) Irreversibility rate, $\dot{I}$.

(iii) $\dot{I}$ when $\dot{Q} = 0$.

Given data:

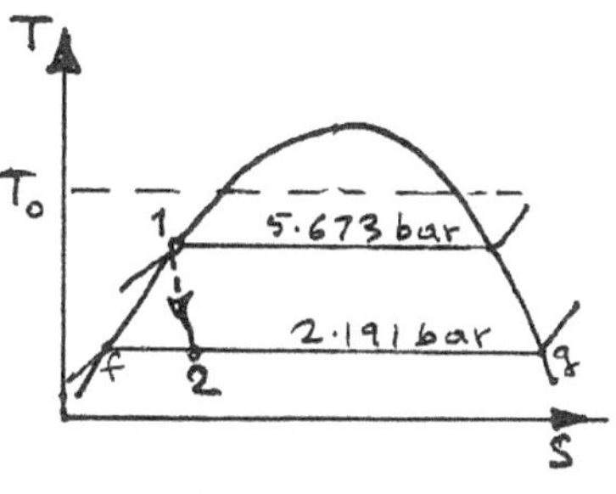

Fig.3.4.b

$\dot{m} = 0.1$ kg/s $\qquad x_2 = 0.21$ when $Q_o \neq 0$.

State 1: sat.liq. at $P_1 = 5.675$ bar

State 2: liq. - vap. region at $P_2 = 2.191$ bar

$T_o = 300$ K

Properties of R-12:

State	h_f kJ/kg	s_f kJ/kgK	h_g kJ/kg	s_g kJ/kgK
1	54.87	0.2078	-	-
2	26.87	0.1080	183.19	0.7020

Analysis:

(i) Heat transfer rate from SFEE, neglecting ΔKE and ΔPE

$$\dot{Q}_o = \dot{m}(\mathrm{h}_2 - h_1)$$

$$h_2 = (1 - x_2)h_f + x_2 h_g$$

$$= 0.79 \times 26.89 + 0.21 \times 18.3.19 = 59.7 \text{ kJ/kg}$$

$$\therefore \dot{Q}_o = 0.1(59.7 - 54.87) = \underline{0.483 \text{ kW}}$$

(ii) Irreversibility rate of the process. The Gouy-Stodola relation

$$\dot{I} = T_o \dot{\Pi} = T_o \left[\dot{m}(s_2 - s_1) - \frac{\dot{Q}_o}{T_o} \right]$$

$$s_2 = (1 - x_2) s_f + x_2 s_g$$

$$= 0.79 \times 0.1080 + 0.21 \times 0.702 = 0.233 \text{ kJ/kg}$$

$$\therefore \dot{I} = 300 \left[0.1(0.233 - 0.2078) - \frac{0.483}{300} \right] = \underline{0.273 \text{ kW}}$$

(iii)Adiabatic throttling

When $\dot{Q}_o = 0$ $\qquad h_2' = h_1 = h_f + (h_g - h_f) x_2'$

$$\therefore x_2 = \frac{h - h_f}{h_g - h_f} = \frac{54.87 - 26.87}{183.19 - 26.87} = 0.179$$

$$s_2' = 0.1080 + (0.702 - 0.108) \times 0.179 = 0.2144 \frac{\text{kJ}}{\text{kg K}}$$

The Gouy-Stodola relation

$$\dot{I}' = T_o \dot{m}(s_2' - s_1) = 300 \times 0.1(0.2144 - 0.2078) = \underline{0.1979 \text{ kW}}$$

Comment: This shows that heat transfer from the environment to a fluid (which is at a temperature $T < T_o$) adds to the irreversibility of the process and therefore good thermal insulation is advisable.

Problem 3.5

The process: Heat transfer form condensing steam to an air stream.

To determine: (i) Heat transfer rate $\dot{Q}_o$.

(ii) Process irreversibility rate.

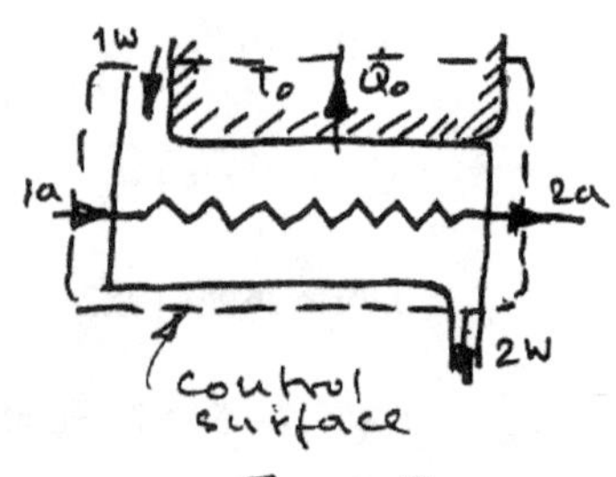

Fig.3.5

Given data:

$\dot{m}_w = 0.5$ kg/s $\qquad \dot{m}_a = 37.5$ kg/s

State 1w: sat.vap.at 0.2 bar

State 2w: sat.liq. at 0.2 bar

For air: $\qquad P_1 = 1.01$ bar $\quad T_1 = 20$ °C

$P_2 = 0.99$ bar $\quad T_2 = 50$ °C $\qquad T_o = 280$ K

Assumptions: Air behaves as a perfect gas.

Analysis:

(1) Heat transfer rate from SFEE, neglecting ΔKE and ΔPE, (assumed -ve)

$$\dot{Q}_o = \dot{m}_w (h_{1w} - h_{2w}) + \dot{m}_a (h_{1a} - h_{2a})$$

H_2O properties

$$h_{1w} - h_{2w} = h_{fg} = 2358 \text{ kJ/kg}$$
$$s_{1w} - s_{2w} = s_{fg} = 7.075 \text{ kJ/kgK}$$
$$\dot{Q}_o = 0.5 \times 2358 + 37.5 \times 1.005(20 - 50)$$

= 54.0 kW to the environment

(ii) Irreversibility rate (from G-S relation)

$$\dot{I} = T_o \left[\dot{m}_w (s_{2w} - s_{1w}) + \dot{m}_a (s_{2a} - s_{1a})\right] + \dot{Q}_o$$

For air $$s_{2a} - s_{1a} = c_P \left[\ln \frac{T_{2a}}{T_{1a}} - \frac{\gamma - 1}{\gamma} \ln \frac{P_{2a}}{P_{1a}} \right]$$

$$= 0.1031 \text{ kJ/kg}$$

$$\therefore \dot{I} = 280[-0.5 \times 7.075 + 37.5 \times 0.1031] + 54.0$$

= 146.05 kW

Comment: There are three phenomena which contribute to this process irreversibility:

1. Pressure losses in the air stream.
2. Heat transfer over finite temperature difference between steam and the air stream.
3. Heat transfer, $\dot{Q}_o$, between the heat exchanger and the environment.

Problem 3.6

The process: Feed water heating by bled steam in a mixing-type feed heater.

To determine: (i) Bled steam mass flow,

(ii) Process irreversibility, both per mass of feed water entering.

Given data:

State C: sat. liq. at 0.05 bar

State S: sat. vap.at 4 bar

State M: sat. liq. at 4 bar

The required thermodynamic properties are:

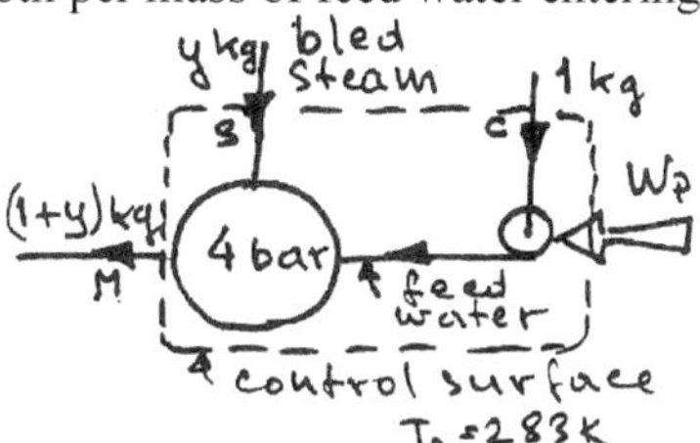

Fig.3.6.a

	$\frac{h}{\text{kJ/kg}}$	$\frac{s}{\text{kJ/kgK}}$
State C	138	0.476
State S	2739	6.897
State M	605	1.776

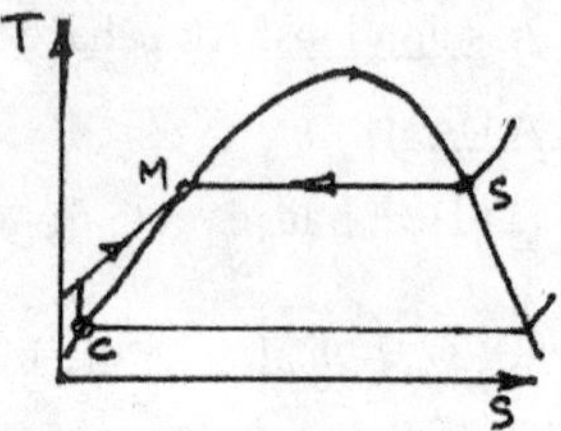

Also, $v_f = 1.078 \times 10^{-3}$ m^3/kg

Fig. 3.6.b

Assumptions: The feed pump process is reversible and adiabatic and water is treated as incompressible.

Analysis:

(i) Bled steam flow rate as fraction of feed water entering, y.

SFEE neglecting ΔKE and ΔPE and $Q = 0$

$$-W = H_{OUT} - H_{IN} \tag{a}$$

Feed pump work $W = -W_p = v_f\,(P_M - P_c)$ (b)

For the control surface shown above

$$v_f\,(P_M - P_c) = h_M\,(1 + y) - (yh_s + h_c) \tag{c}$$

$$\therefore y = \frac{h_M - h_c - v_f\,(P_M - P_c)}{h_s - h_M} \tag{d}$$

Substituting

$$y = \frac{605 - 138 - 1.078 \times 10^{-3}\,(4 - 0.05) \times 10^5 \times 10^{-3}}{2739 - 605}$$

$= \underline{0.219}$

(ii) Process irreversibility per mass of feed water entering.

Using the Gouy-Stodola relation for an adiabatic system

$I = T_o\,(s_{OUT} - s_{IN})$

$= T_o[(1+y)s_M - s_c - ys_s]$

substituting

$I = 283\,[(1+0.219) \times 1.776 - 0.476 + 0.219 \times 6.897]$

$= \underline{49.2 \text{ kJ}}$ per kg of feed water entering.

Comment: The irreversibility of this process is due to mixing of two streams of H_2O in different thermodynamic states.

Problem 3.7

The process: Mixing of two air streams.

$\dot{m}_A = 3$ kg/s, A, MIXING PROCESS, $\dot{m}_C$, C, B, $\dot{m}_B = 2$ kg/s, $Q = 0$

Fig. 3.7.a

To determine: (i) Exergy flow rates of the three steams involved.

(ii) To calculate, hence, the irreversibility rate of the mixing process.

Given data:

Data as in Problem 1.6 with $T_o = 300$ K $\quad P_o = 1$ bar

Assumptions: The process is adiabatic, air behaves as a perfect gas.

Analysis:

(i) Since for air $\varepsilon_o = 0$

$$\dot{E}_1 = \dot{E}_{\text{ph},1} = \dot{m}[(h_1 - h_o) - T_o(s_1 - s_o)] \tag{a}$$

For a perfect gas

$$\dot{E}_1 = \dot{m}c_P\left[(T_1 - T_o) - T_o \ln\frac{T_1}{T_o} + \frac{\gamma - 1}{\gamma} T_o \ln\frac{P_1}{P_o}\right] \tag{b}$$

Substituting numerical values from Problem 1.6

$\dot{E}_A = 145.4$ kW $\qquad \dot{E}_B = 30.8$ kW $\qquad \dot{E}_C = 154.5$ kW

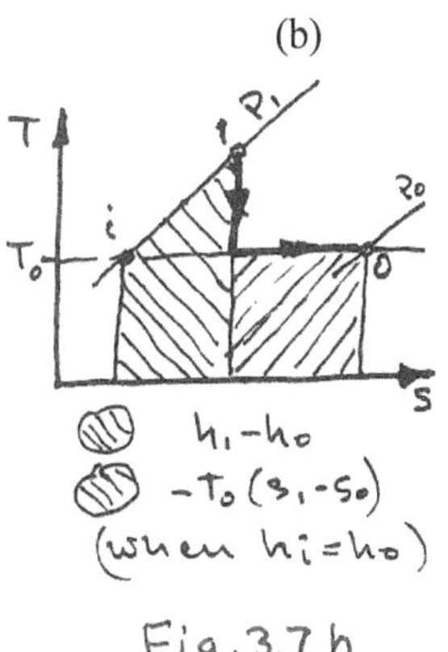

Fig. 3.7.b

(ii) The exergy balance with $\dot{Q} = 0 \quad \dot{W}_x = 0$

$$\dot{I} = \dot{E}_A + \dot{E}_B - \dot{E}_C$$

$= 145.4 + 30.8 - 154.5 = \underline{21.7 \text{ kW}}$

Using the Gouy-Stodola relation and the value of $\dot{\Pi}$ from problem 1.6

$$\dot{I} = T_o\dot{\Pi} = 300 \times 0.072 = 21.6 \text{ kW}$$

The difference is due to rounding error.

Problem 3.8

Process: Heat transfer from the environment to a stream of refrigerated brine.

To determine: The irreversibility rate of the process and represent it on a *T-S* diagram.

Given data:

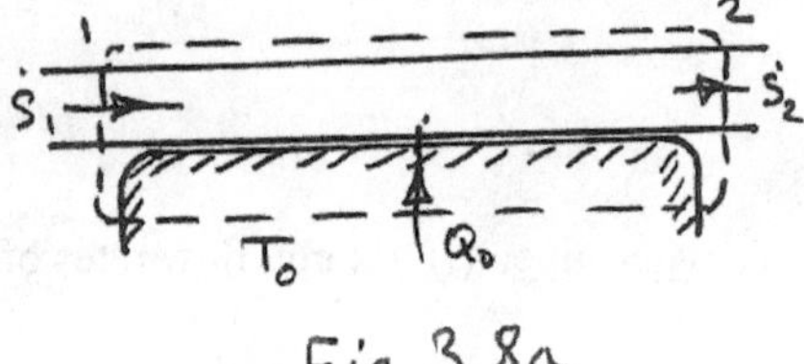

Fig. 3.8a

$\dot{m} = 0.5$ kg/s $\qquad T_1 = 240$ K

$T_2 = 243$ K $\qquad T_o = 300$ K

For brine: $c_P = 2.85$ kJ/kgK

Assumptions: Pressure losses are negligible.

Analysis: Using the Gouy-Stodola relation with $\dot{Q}_i = \dot{Q}_o$ and $T_i = T_o$

$$\dot{I} = T_o \Pi = T_o\left[\left(\dot{S}_2 - \dot{S}_1\right) - \frac{\dot{Q}_o}{T_o}\right] \qquad \text{(a)}$$

From the energy balance

$$\dot{Q}_o = \dot{m}_b c_{Pb}(T_2 - T_1) \qquad \text{(b)}$$

$$= 0.5 \times 2.85\,(243 - 240) = 4.275 \text{ kW}$$

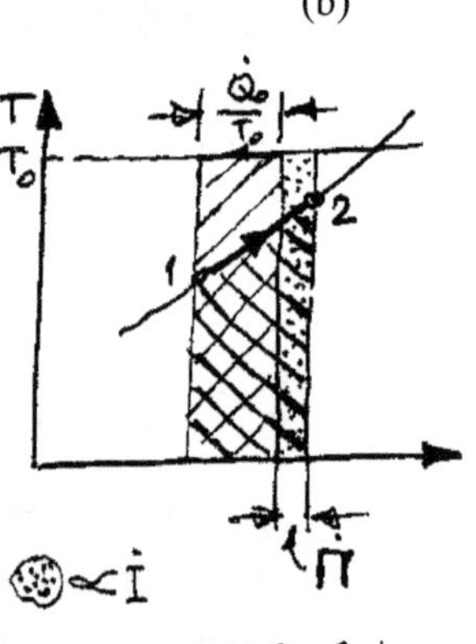

Fig. 3.8.b

Since P = const

$$\dot{S}_2 - \dot{S}_1 = \dot{m}_b c_{Pb} \ln\frac{T_2}{T_1} = 0.0177 \text{ kW/K}$$

Substituting in (a)

$$\dot{I} = 300 \times 0.0177 - 4.275$$

$$= \underline{1.035 \text{ kW}}$$

The T-$\dot{S}$ diagram interprets eq. (a). N.B. The two cross-hatched areas are equal.

Problem 3.9

Process: De-superheating of steam.

To determine: (i) Mass flow rate of water.
(ii) Irreversibility rate of process.

Given data:

State 1: 400 °C, 20 bar, 0.5 kg/s

State 2: sat. vapour at 20 bar

State 3: sat. liquid at 20 bar

Hence, properties:

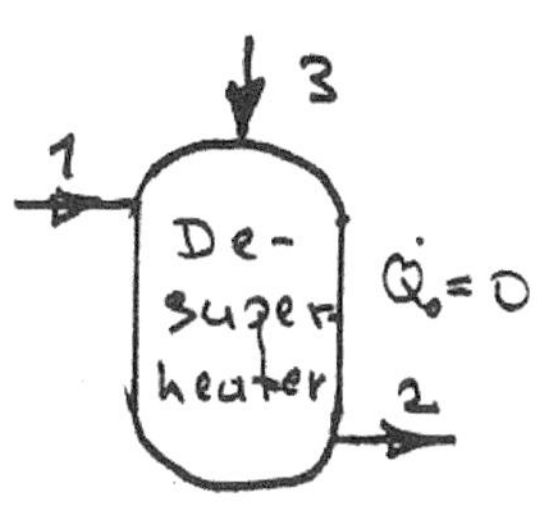

Fig. 3.9

	h kJ/kg	s kJ/kg K
State 1	3248	7.126
State 2	2799	6.340
State 3	909	2.447

Assumptions: The process is adiabatic.

Analysis: using the Gouy-Stodola relation with $Q = 0$

$$\dot{I} = T_o \dot{\Pi} = (\dot{S}_{OUT} - \dot{S}_{IN})T_o$$

$$= (\dot{m}_2 s_2 - \dot{m}_1 s_1 - \dot{m}_3 s_3)T_o \qquad \text{(a)}$$

Energy balance $\quad (\dot{Q}_o = 0,\ \Delta KE = 0,\ \Delta PE = 0)$

$$\dot{m}_1 h_1 + \dot{m}_3 h_3 = m_2 h_2 \qquad \text{(b)}$$

But $\dot{m}_2 = \dot{m}_1 + \dot{m}_3 \qquad$ (c)

$$\therefore \dot{m}_3 = \frac{m_1(h_1 - h_2)}{h_2 - h_3} = \frac{0.5(3248 - 2799)}{2799 - 909} = \underline{0.119 \text{ kg/s}}$$

Substituting in (a)

$$\dot{I} = 300\,(0.619 \times 6.340 - 0.5 \times 7.126 - 0.119 \times 2.447)$$

$$= \underline{21.08 \text{ kW}}$$

Problem 3.10

Process: Heat exchange in an evaporator between ammonia and brine.

To determine: (i) Stray heat transfer rate between the evaporator and its surroundings.

(ii) Irreversibility rate of the process.

Given data:

$\dot{m}_a = 0.45$ kg/s $\qquad \dot{m}_b = 16$ kg/s

$P_a = 1.902$ bar $\qquad x_{a1} = 0.25$

$(\Delta T_{su})_{a2} = 10$ K $\qquad c_{Pb} = 2.85$ kJ/kgK

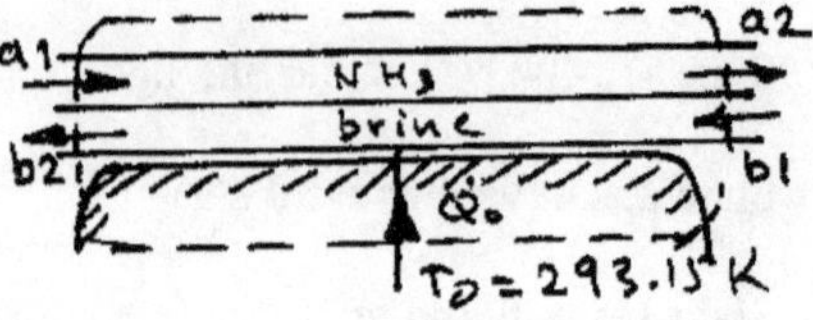

Fig.3.10.a

From NH_3 tables:

$h_{a1} = 89.8 + 0.25\,(1420 - 89.8) = 422.35$ kJ/kg

$s_{a1} = 0.368 + 0.25\,(5.623 - 0.368) = 1.682$ kJ/kgK

$$\left.\begin{array}{l} h_{a2} = 1442.96 \text{ kJ/kg} \\ s_{a2} = 5.7062 \text{ kJ/kgK} \end{array}\right\} \text{by interpolation}$$

Assumptions: $\Delta P = 0,\ \ \Delta KE = 0,\ \ \Delta PE = 0\ \ c_{Pb} = \text{const}$

Analysis:

(i) Energy balance for the control region

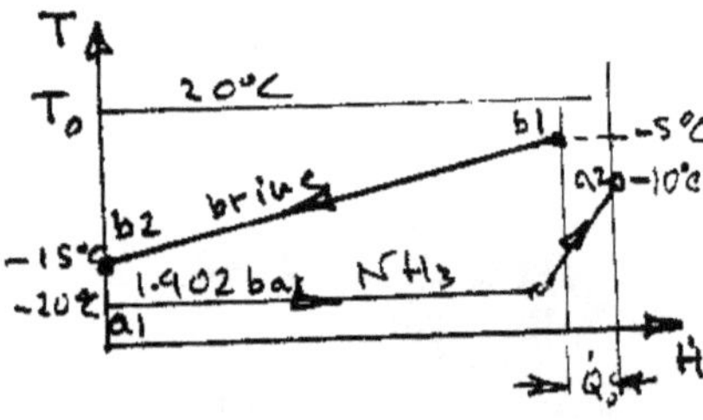

Fig.3.10.b

$$\dot{Q}_o = m_b c_{Pb}(T_{b2} - T_{b1}) + \dot{m}_a(h_{a2} - h_{a1})$$

$$= \underline{3.27 \text{ kW}}$$

(ii) From the Gouy-Stodola relation $(\dot{Q} = \dot{Q}_o)$

$$\dot{I} = T_o\left[\dot{m}_a(s_{a2} - s_{a1}) + \dot{m}_b(s_{b2} - s_{b1})\right] - \dot{Q}_o$$

$$= 293.15\left[0.45(5.7062 - 1.682) + 16 \times 2.85 \ln\frac{258.15}{268.15}\right] - 3.27$$

$$= \underline{19.6 \text{ kW}}$$

Problem 3.11

Process: Heat exchange between condensing mercury and evaporating water substance.

To determine: (i) Thermal exergy flow rates at T_M and T_w,

(ii) Process irreversibility rate.

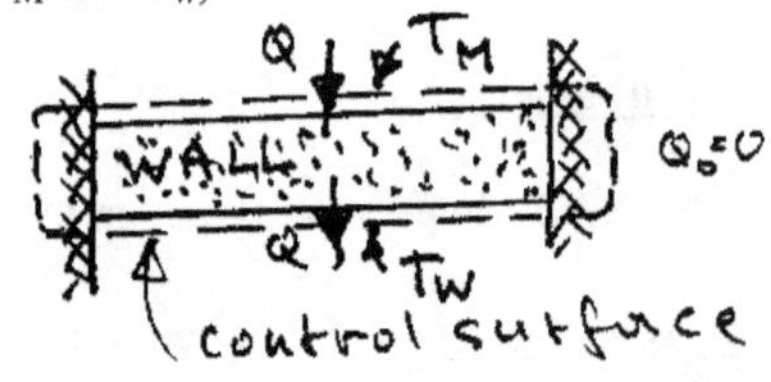

Fig.3-11

Given data:

$\dot{Q} = 500$ MW $\qquad T_o = 300$ K

$T_M = 580$ K $\qquad T_w = 500$ K

Assumptions: Pressure losses and heat losses are negligible.

Analysis:

(i) For the control surface shown

$$\dot{E}^Q_M = \dot{Q}\frac{T_M - T_o}{T_M} = 500\text{ MW}\frac{590 - 300}{590} = 245.8\text{ MW}$$

$$\dot{E}^Q_M = -\dot{Q}\frac{T_W - T_o}{T_W} = -500\text{ MW}\frac{550 - 300}{550} = -227.3\text{ MW}$$

(ii) Exergy balance for the control surface shown $\quad (\dot{E}_1 = 0,\ \dot{E}_2 = 0)$,

$$\dot{E}_1 + \sum_i \dot{E}^Q_i = \dot{E}_2 + \dot{W}_x + \dot{I}$$

$$\therefore\ \sum_i \dot{E}^Q_i = \dot{E}^Q_M + \dot{E}^Q_W$$

$$= 245.8 - 227.3 = \underline{18.5\text{ MW}}$$

(iii)Using the value of $\dot{\Pi}$ calculated in Problem 1.9 ($\dot{\Pi} = 61.6$ kW/K)

$$\dot{I} = T_o\dot{\Pi} = 300\text{ K}\times 61.6\text{ kW/K} = \underline{18480\text{ kW}}$$

Comment: the same answer is obtained.

Problem 3.12

Derivation: Ψ for the plant.

Exergy balance

$$\dot{E}_i + \dot{E}^Q = \dot{E}_e + \dot{W}_x + \dot{I} \qquad \text{(a)}$$

But for this plant

$$\dot{E}_i = 0 \quad \dot{E}_e = 0 \qquad \text{(b)}$$

Hence,

$$\dot{E}^Q = \dot{W}_x + \dot{I} \qquad \text{(c)}$$

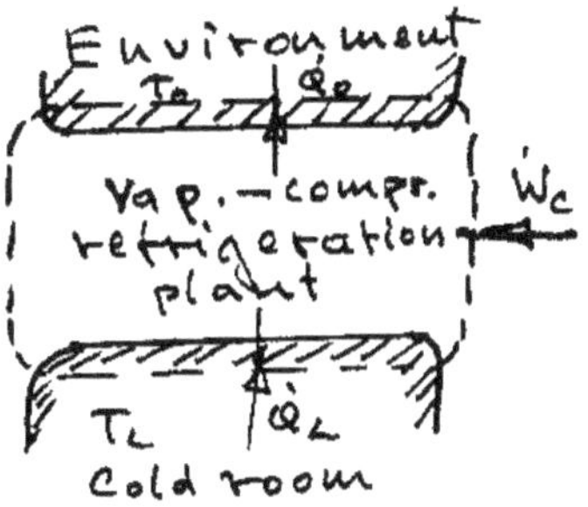

Fig. 3.12.a

where $\dot{E}^{Q} = \dot{Q}_L \dfrac{T_L - T_o}{T_L} = -\dot{Q}_L \dfrac{T_o - T_L}{T_L}; \dot{W}_x = -\dot{W}_c$

Substituting in (c)

$$\underbrace{\dot{Q}_L \frac{T_o - T_L}{T_L}}_{\text{OUTPUT}} = \underbrace{\dot{W}_c - \dot{I}}_{\text{INPUT}} \tag{d}$$

Hence, rational (or exergetic) efficiency

$$\Psi = \frac{\text{OUTPUT}}{\text{INPUT}} = \frac{\dot{Q}_L / \dot{W}_c}{T_L / (T_o - T_L)} = \frac{(\text{CP})_{\text{REF}}}{(\text{CP})_{\text{CARNOT}}} \tag{e}$$

From (d) and (e) an alternative expression is obtained

$$\underline{\Psi = 1 - \frac{\dot{I}}{\dot{W}_c}} \tag{f}$$

<u>Process</u>: Ammonia, vapour-compression refrigeration plant.

<u>To determine</u>:

(i) Specific irreversibilities and corresponding efficiency defects of the components,
(ii) the rational efficiency of the plant,
(iii)construct a pie diagram for the plant

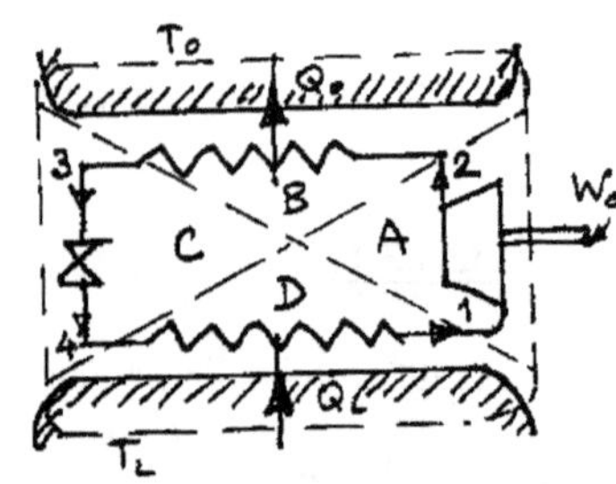

Fig.3.12.b

Given data:

Compressor efficiency $\eta_s = 0.73$

Cycle parameters as shown.

From NH_3 tables:

$h_1 = 1433.0$ kJ/kg $\qquad s_1 = 5.475$ kJ/kgK

$h_3 = 323.1$ kJ/kg $\qquad s_3 = 1.204$ kJ/kgK

$h_4 = h_3 = 323.1$ kJ/kg

$$s_4 = s_g - \frac{h_g - h_4}{T_{\text{EVAP}}} = 5.475 - \frac{1433 - 323.1}{263.15} = 1.257 \text{ kJ/kgK}$$

Since $s_{2'} = s_1$, by interpolation, $h_{2'} = 1632.3$ kJ/kg

Hence, $h_2 = h_1 + \dfrac{h_{2'} - h_1}{\eta_s} = 1706.0$ kJ/kg

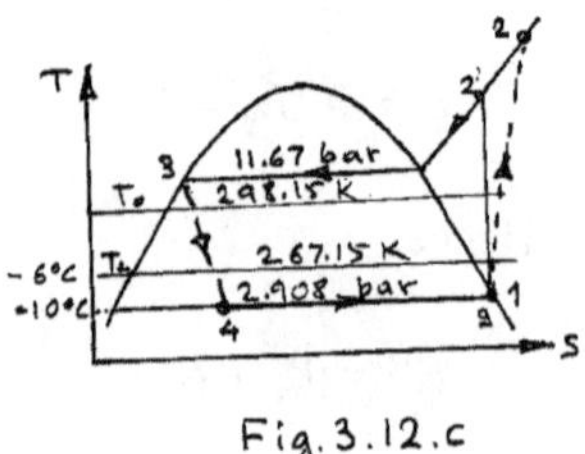

Fig.3.12.c

Assumptions: The valve and the compressor are adiabatic, pressure losses as well as ΔKE and ΔPE are negligible.

Analysis:

(i) Gouy-Stodola relation

$$\dot{I} = T_o\left[(\dot{s}_{OUT} - \dot{s}_{IN}) - \sum \frac{\dot{Q}_i}{T_i}\right]$$

Specific irreversibility $\quad i = \dot{I} / \dot{m}$

Efficiency defect of the k-th plant component $\quad \delta_K = i_c / W_c$

Interactions:

$$q_L = \frac{\dot{Q}_L}{\dot{m}} = 1433 - 323.1 = 1109.9 \text{ kJ/kg}$$

$$q_o = \frac{\dot{Q}_o}{\dot{m}} = 1706 - 323.1 = 1382.9 \text{ kJ/kg}$$

$$W_c = \frac{\dot{W}_c}{\dot{m}} = 1706 - 1433 = 273.0 \text{ kJ/kg}$$

Subregion A - compressor ($1 \rightarrow 2$)

$q_{1,2} = 0 \therefore i_A = T_o (s_2 - s_1) = \underline{58.44 \text{ kJ/kg}}$

$$\delta_A = \frac{58.44}{273} = \underline{0.214}$$

Subregion B - condenser ($2 \rightarrow 3$)

$i_B = T_o (s_3 - s_2) + q_o = \underline{51.1 \text{ kJ/kg}}$

$$\delta_B = \frac{51.1}{273} = \underline{0.187}$$

Subregion C - Throttling valve ($3 \rightarrow 4$)

$Q_{3,4} = 0 \therefore i_C = T_o (s_4 - s_3) = \underline{15.8 \text{ kJ/kg}}$

$$\delta_C = \frac{15.8}{273} = \underline{0.058}$$

Subregion D - evaporator (4 → 1)

$$I_D = T_o\left[(s_1 - s_4) - \frac{q_L}{T_L}\right] = \underline{18.9\ \text{kJ/kg}}$$

$$\delta_D = \frac{18.9}{273} = \underline{0.069}$$

(ii) Rational efficiency - using eq. (e)

$$\psi = \frac{1109.9}{273} \bigg/ \frac{267.15}{31} = \underline{0.472} \quad (\text{Also, } \psi = 1 - \sum_k \delta_K\ )$$

(iii)Pie diagram

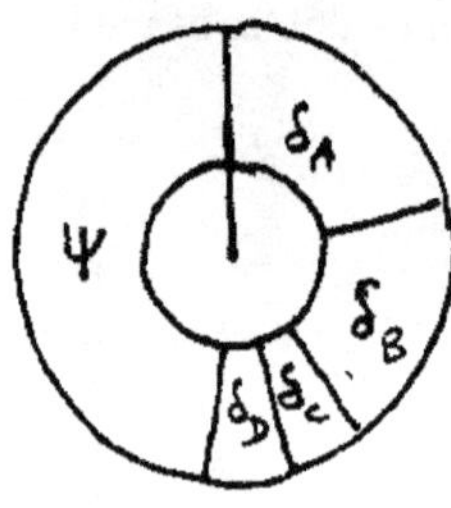

$$\theta_A = 0.214 \times 360° = 77.1°$$

$$\theta_B = 0.187 \times 360° = 67.3°$$

$$\theta_C = 0.058 \times 360° = 20.8°$$

$$\theta_D = 0.069 \times 360° = 24.9°$$

Fig. 3.12.d

Problem 3.13

Process: Heat exchange between NH_3 and air in a heat-pump condenser.

To determine:
(i) electric power input to the fan,
(ii) mass flow rate of the NH_3 stream,
(iii)rational efficiency of the system.

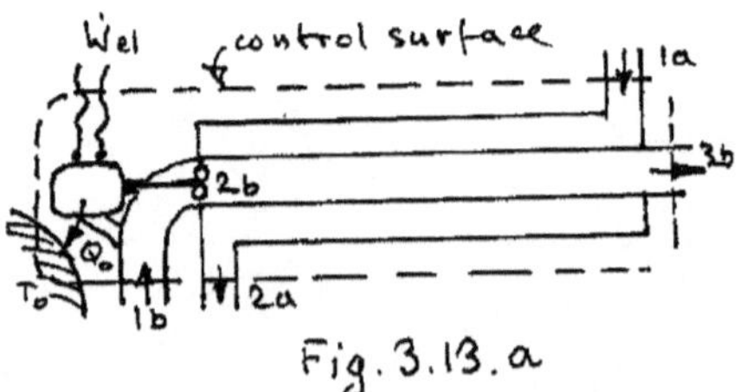

Fig. 3.13.a

Given data:

$\dot{m}_b = 0.5$ kg/s $P_a = 15.54$ bar
$P_{b1} = P_{b3} = 1$ bar State 2a: sat.liq. at P_a
$P_{2a} = 1.005$ bar Motor overall eff. $\eta_{ov} = 0.7$
Fan isentropic eff. $\eta_s = 0.8$
$T_o = 280$ K

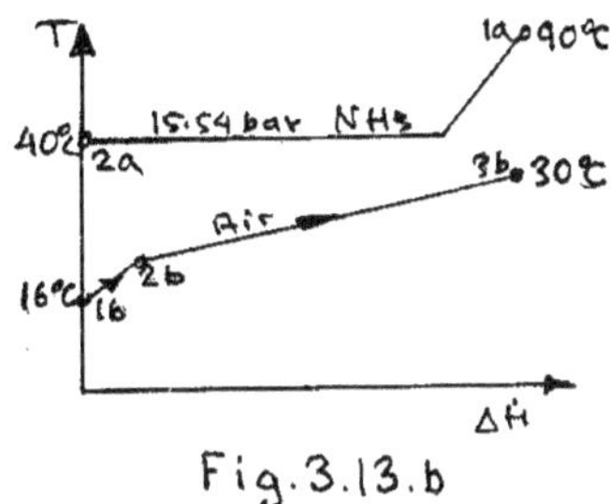

Fig.3.13.b

NH_3 properties:

$h_{1a} = 1622.4$ kJ/kg $\qquad h_{2a} = 371.9$ kJ/kg,

$s_{1a} = 5.321$ kJ/kgK $\qquad s_{2a} = 1.360$ kJ/kgK

Assumptions: No pressure losses in NH_3 stream, negligible ΔKE and ΔPE. Air behaves as a perfect gas.
Analysis:

(i) Electric power input to the fan

$$\dot{W}_{el} = \dot{m}_b c_{Pb} T_{1b} \left[\left(\frac{P_{2b}}{P_{1b}} \right)^{\frac{\gamma-1}{\gamma}} - 1 \right] \Big/ (\eta_{ov} \eta_s) \qquad \text{(a)}$$

$$= 0.5 \times 1.005 \times 289.15 \left[\left(\frac{1.005}{1.00} \right)^{\frac{0.4}{1.4}} - 1 \right] \Big/ (0.7 \times 0.8)$$

$$= \underline{0.37 \text{ kW}}$$

(ii) Mass-flow rate of the NH_3 stream

Energy balance for the system

$$-\dot{Q}_o + W_{el} = \dot{m}_b c_{Pb} (T_{3b} - T_{1b}) + \dot{m}_a (h_{2a} - h_{1a}) \qquad \text{(b)}$$

$$\therefore \ \dot{m}_a = \frac{\dot{m}_b c_{pb} (T_{3b} - T_{1b}) - \dot{W}_{el} \times \eta_{ov}}{h_{1a} - h_{2a}} = \underline{5.42 \times 10^{-3} \text{ kg/s}}$$

(iii) The rational efficiency of the system

Exergy balance for the system:

$$\underbrace{\left(\dot{E}_{1a} - \dot{E}_{2a}\right) + \dot{W}_{el}}_{\text{INPUT}} - \underbrace{\left(\dot{E}_{3b} - \dot{E}_{1b}\right)}_{\text{OUTPUT}} = \dot{I}$$

Rational efficiency is defined

$$\Psi = \frac{\dot{E}_{3b} - \dot{E}_{1b}}{(\dot{E}_{1a} - \dot{E}_{2a}) + \dot{W}_{el}} \qquad \text{(c)}$$

$$\dot{E}_{1a} - \dot{E}_{2a} = \dot{m}_a \left[(h_{1a} - h_{2a}) - T_o (s_{1a} - s_{2a}) \right]$$

$$= 0.766 \text{ kW}$$

Since $P_{1b} = P_{3b}$

$$\dot{E}_{3b} - \dot{E}_{1b} = \dot{m}_b c_P \left[(T_{3b} - T_{1b}) - T_o \ln \frac{T_{3b}}{T_{1b}} \right]$$

$$= 0.382 \text{ kW}$$

Substituting in (c)

$$\Psi = \frac{0.382}{0.766 + 0.37} = \underline{0.337}$$

Comment: Formulation of the rational efficiency is facilitated by inspecting the exergy balance for the system and identifying in it the input and the output.

Problem 3.14

Process: Heat pump operating on an open, reversed Joule, air cycle.

To determine: (i) specific irreversibilities of the components,
(ii) external irreversibility due to mixing
(iii) rational efficiency.

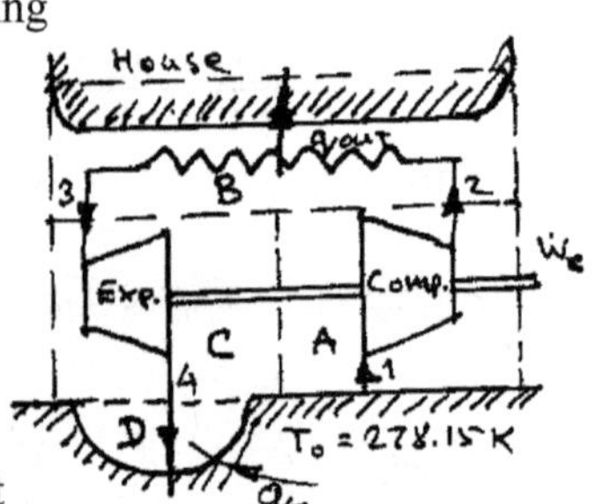

Fig.3.14.a

Given data:

$$\eta_{exp} = \eta_{comp} = 0.80$$
$$T_3 = 305.15 \text{ K}$$
$$P_3 = 1.96 \text{ bar}$$

Assumptions: The air is dry and behaves as a perfect gas, the compressor and the expander are adiabatic.

Analysis: Cycle analysis:

$$\frac{T_2'}{T_1} = \left(\frac{P_2}{P_1} \right)^{\frac{\gamma - 1}{\gamma}} \qquad T_2' = 278.15 \left(\frac{2}{0.98} \right)^{\frac{0.4}{1.4}} = 341.0 \text{ K}$$

$$T_2 = T_1 + \frac{1}{\eta_c}\left(T_2^{'} - T_1\right) = 278.15 + \frac{1}{0.8}(341.0 - 278.15)$$

$$= \underline{356.7 \text{ K}}$$

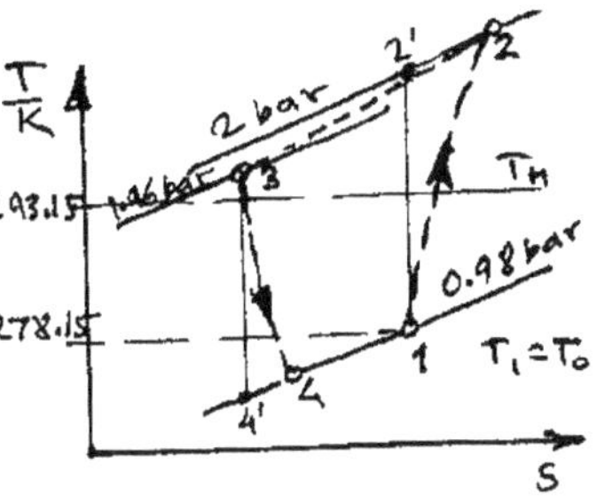

Similarly

$$\frac{T_4^{'}}{T_3} = \left(\frac{P_4}{P_3}\right)^{\frac{\gamma-1}{\gamma}} \quad \therefore \; T_4^{'} = 250.3 \text{ K}$$

$$T_4 = T_3 - \eta_T (T_3 - T_{4'}) = \underline{261.3 \text{ K}}$$

Fig.3.14.b

Hence,

$$-q_{\text{OUT}} = q_{2,3} = c_P (T_3 - T_2) = 1.005 \; (305.15 - 356.7)$$

$$= \underline{-51.56 \text{ kJ/kg}}$$

$$w_c = \frac{\dot{W}_c}{\dot{m}} = c_P\left[(T_2 - T_1) - (T_3 - T_4)\right]$$

$$= 1.005 \; [(356.7 - 278.15) - (305.15 - 261.3)]$$

$$= \underline{34.87 \text{ kJ/kg}}$$

Heat transfer to the mixing region

$$q_o = c_P (T_o - T_4) = \underline{16.93 \text{ kJ/kg}}$$

(i) Component irreversibilities

Subregion A - Compressor ($1 \rightarrow 2$)

Since $q = 0$, i_A - $T_o(s_2 - s_1)$

$$= T_o c_P \left(\ln\frac{T_2}{T_1} - \frac{\gamma - 1}{\gamma} \ln\frac{P_2}{P_1} \right)$$

$$= \underline{12.55 \text{ kJ/kg}}$$

Subregion B - Heat Exchanger ($2 \rightarrow 3$)

$$I_B = T_o\left[(s_3 - s_2) + \frac{q_{\text{OUT}}}{T_h}\right]$$

$$= T_o\left[c_P\left(\ln\frac{T_3}{T_2} - \frac{\gamma-1}{\gamma}\ln\frac{P_3}{P_2}\right) + \frac{q_{OUT}}{T_H}\right]$$

$= \underline{7.155 \text{ kJ/kg}}$

Subregion C - Expander ($3 \to 4$)

Since $q = 0$, $i_c = T_o(s_4 - s_3)$

$$= T_o c_P\left(\ln\frac{T_4}{T_3} - \frac{\gamma-1}{\gamma}\ln\frac{P_4}{P_3}\right)$$

$= \underline{11.99 \text{ kJ/kg}}$

(ii) Subregion D - Mixing ($4 \to 0$)

i_c - $T_o(s_o - s_4) - q_o$

$$= T_o c_P \ln\frac{T_o}{T_4} - q_o = \underline{0.535 \text{ kJ/kg}}$$

(iii) Rational efficiency

Total irreversibility $\quad i_T = \sum_k i_k = 32.23$ kJ/kg

Hence, $\Psi = 1 - \frac{\sum_k i}{w_c} = 1 - \frac{32.23}{34.87} = \underline{0.0758}$

Comment: Note the low plant rational efficiency.

Problem 3.15

Process: A single stage reciprocating compressor with an aftercooler.

To determine: (i) The interactions $\dot{W}_{xc}$ $\dot{Q}_c$ and $\dot{Q}_a$, and

(ii) construct a Grassmann diagram.

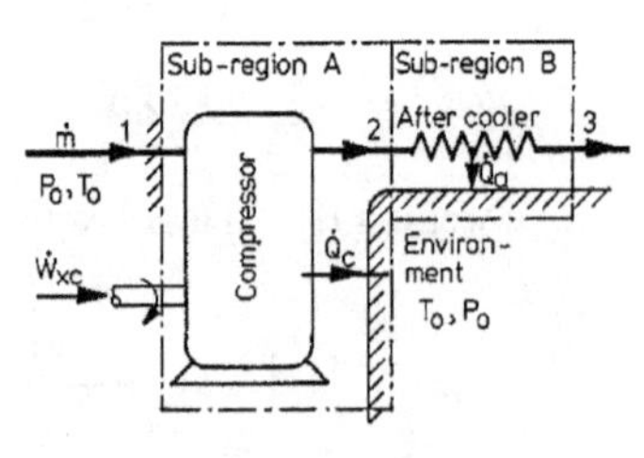

Fig. 3.15.a

Given data:

$\dot{m} = 0.1$ kg/s	$\eta_{iso} = 0.68$
$T_1 = T_o = 288$ K	$P_1 = P_o = 1$ bar
$P_2 = 4$ bar	$T_2 = 400$ K
$T_3 = 293$ K	$P_3 = 3.90$ bar

Assumptions: Quasi-steady operation of the plant, negligible ΔKE and ΔPE, air behaves as a perfect gas.

Analysis:

(i) Energy transfer rates

Reversible isothermal compression

$$\dot{W}_{iso} = \dot{m}RT_o \ln\frac{P_2}{P_1} = 0.1 \times 0.287 \times 288 \ln 4 = 11.46\,\text{kW}$$

Shaft power input

$$\dot{W}_{xc} = \dot{W}_{iso} / \eta_{iso} = 11.46 / 0.68 = \underline{16.85\,\text{kW}}$$

Cylinder heat transfer rate

$$-\dot{Q}_c = -\dot{W}_{xc} + \dot{m}c_P(T_2 - T_1)$$

$$= -16.85 + 0.1 \times 1.005\,(400 - 288) = \underline{-5.60\,\text{kW}}$$

After cooler heat transfer rate

$$-\dot{Q}_a = \dot{m}c_P(T_3 - T_2)$$

$$= 0.1 \times 1.005\,(293 - 400) = \underline{-10.75\,\text{kW}}$$

(ii) Exergy flow rates

For an air stream $(\varepsilon_o = 0)$

$$\dot{E} = \dot{m}[(h - h_o) - T_o(s - s_o)] \qquad \text{(a)}$$

When regarded as a perfect gas

$$\dot{E} = \dot{m}R\left\{\frac{\gamma}{\gamma - 1}\left[(T - T_o) - T_o \ln\frac{T}{T_o}\right] + T_o \ln\frac{P}{P_o}\right\} \qquad \text{(b)}$$

State 1

Since $P_1 = P_o$, $T_1 = T_o$ $\quad \therefore \underline{\dot{E}_1 = 0}$

State 2

$$\dot{E}_2 = 01 \times 0.287\left\{3.5\left[(400 - 288) - 288 \ln\frac{400}{288}\right] + 288 \ln 4\right\}$$

= $\underline{13.21\text{ kW}}$

State 3

$$\dot{E}_3 = 01 \times 0.287\left\{3.5\left[(293-288) - 288\ln\frac{293}{288}\right] + 288\ln 3.9\right\}$$

= $\underline{11.25\text{kW}}$

Thermal exergy flows

In general $\quad \dot{E}^{\mathrm{Q}} = \dot{Q}_{\mathrm{R}}\dfrac{T_{\mathrm{R}} - T_{\mathrm{o}}}{T_{\mathrm{R}}}$ (c)

But since both the heat transfers $\dot{Q}_{\mathrm{c}}$ and $\dot{Q}_{\mathrm{a}}$ are ultimately to the environment at $T_{\mathrm{R}} = T_{\mathrm{o}}$, we have

$$\underline{E_{\mathrm{c}}^{\mathrm{Q}} = 0} \text{ and } \underline{\dot{E}_{\mathrm{a}}^{\mathrm{Q}} = 0}$$

Irreversibilities

Exergy balance $\quad \dot{E}_{\mathrm{IN}} + E^{\mathrm{Q}} = \dot{E}_{\mathrm{OUT}} + \dot{W}_{\mathrm{x}} + \dot{I}$ (d)

Sub region A - with $\quad \dot{E}_1 = 0,\ \dot{E}_{\mathrm{c}}^{\mathrm{Q}} = 0,\ \dot{W}_{\mathrm{x}} = -\dot{W}_{\mathrm{xc}}$

we get

$$\dot{I}_{\mathrm{A}} = \dot{W}_{\mathrm{xc}} - \dot{E}_2 = 16.85 - 13.21 = \underline{3.64\text{ kW}}$$

Sub region B - with $\quad \dot{E}_{\mathrm{a}}^{\mathrm{Q}} = 0,\ \dot{W}_{\mathrm{x}} = 0$

we get

$$\dot{I}_{\mathrm{B}} = \dot{E}_2 - \dot{E}_3 \qquad \text{(e)}$$

Using perfect gas relations in (e)

$$\dot{I}_{\mathrm{B}} = \dot{m}R\left\{\frac{\gamma}{\gamma-1}\left[(T_2 - T_3) - T_{\mathrm{o}}\ln\frac{T_2}{T_3}\right] + T_{\mathrm{o}}\ln\frac{P_2}{P_3}\right\} \qquad \text{(f)}$$

$$= \dot{I}_{\mathrm{B}}^{\Delta T} + \dot{I}_{\mathrm{B}}^{\Delta P}$$

where $\dot{I}_{\mathrm{B}}^{\Delta T} = \dot{m}R\dfrac{\gamma}{\gamma-1}\left[(T_2 - T_3) - T_{\mathrm{o}}\ln\dfrac{T_2}{T_3}\right]$ (g)

$$\dot{I}_{\mathrm{B}}^{\Delta P} = \dot{m}RT_{\mathrm{o}}\ln\frac{P_2}{P_3} \qquad \text{(h)}$$

Substituting numerical values in (g) and (h)

$\dot{I}_{B}^{\Delta T} = \underline{1.74 \text{ kW}}$ and $\dot{I}_{B}^{\Delta P} = \underline{0.21 \text{ kW}}$

Hence, $\dot{I}_{B} = 1.74 + 0.21 = \underline{1.95 \text{ kW}}$

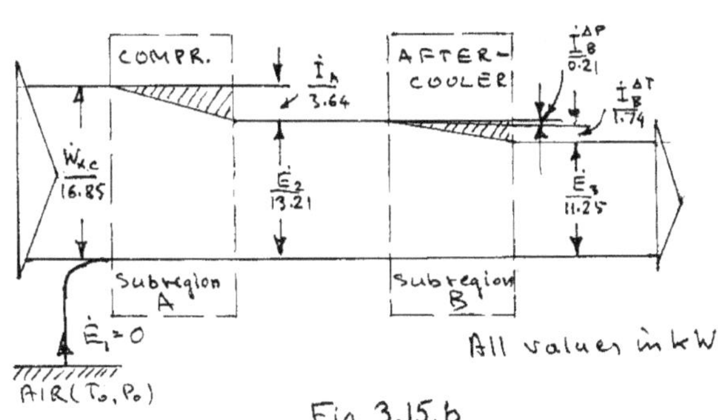

Fig.3.15.b

Problem 3.16

Process: Maintaining a cold chamber at a constant temperature with the aid of a vapour-compression refrigerator.

To determine: The irreversibility rate attributable to electric power dissipation of 500 W in the cold chamber when the latter is to be maintained at a temperature of:

(i) 260 K, (ii) 240 K.

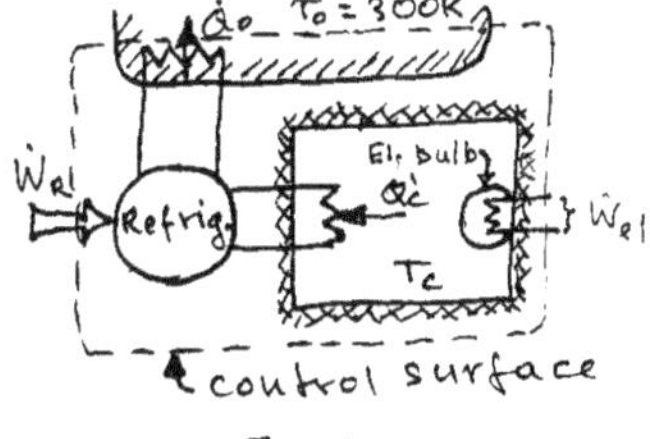

Fig.3.16

Given data: $\Psi_{REF} = 0.4$ $\dot{W}_{el} = 500$ W
(i) $T_c = 260$ K
(ii) $T_c = 240$ K

Assumptions: The cold chamber is adiabatic. ($Q_o = 0$).

Analysis: The Gouy-Stodola relation for the control surface shown

$$\dot{I} = T_o\left[\left(\dot{S}_{OUT} - \dot{S}_{IN}\right) - \sum \frac{\dot{Q}_i}{T_i}\right]$$

But, $\dot{S}_{OUT} = \dot{S}_{IN} = 0,\ \dot{Q}_i = -\dot{Q}_o \quad T_i = T_o$

Hence, $\dot{I} = \dot{Q}_o$ (1)

For the refrigerator $\dot{W}_R = \dot{Q}_o - \dot{Q}_c$ (2)

Rational efficiency $\quad \Psi_{\text{REF}} = \dfrac{(\text{CP})_{\text{REF}}}{(\text{CP})_{\text{CARNOT}}}$ (3)

$$\Psi_{\text{REF}} = \frac{\dot{Q}_{\text{c}}/\dot{W}_{\text{R}}}{T_{\text{c}}/(T_{\text{o}} - T_{\text{c}})} \tag{4}$$

Using (2) in (4) and re-arranging, with $\dot{Q}_{\text{c}} = \dot{W}_{\text{el}}$

$$\dot{Q}_{\text{o}} = \dot{W}_{\text{el}}\left(\frac{T_{\text{o}} - T_{\text{c}}}{\Psi_{\text{REF}} T_{\text{c}}} + 1\right) \tag{5}$$

Case (1) $\quad T_{\text{c}} = 260$ K

$$\dot{I} = \dot{Q}_{\text{o}} = 500 \text{ W}\left(\frac{300 - 260}{0.4 \times 260} + 1\right)$$
$$= \underline{692.3 \text{ W}}$$

Case (ii) $\quad T_{\text{c}} = 240$ W

$$\dot{I} = 500 \text{ W}\left(\frac{300 - 240}{0.4 \times 240} + 1\right)$$

$$= \underline{812.5 \text{ W}}$$

Comment: The lower is the temperature at which high grade energy is dissipated (say, through ohmic or viscous friction, magnetic or mechanical hysteresis etc.) the higher is the total irreversibility attributable to it. (N.B. This applies to $T < T_{\text{o}}$).

Problem 3.17

Process: A proposed engine running on fuel burnt at a maximum temperature of 1000 K and using a heat sink maintained at 200 K by a refrigerator driven by the engine.

To determine: Thermodynamic feasibility of the engine.

Given data: $\dot{W}_{\text{NET}} = 720$ kW $\qquad \Delta h_{\text{o}} = 40000$ kJ/kg

$\dot{m}_{\text{f}} = 0.025$ kg/s $\qquad T_{\text{H}} = 1000$ K

$T_{\text{L}} = 200$ K $\qquad T_{\text{o}} = 300$ K

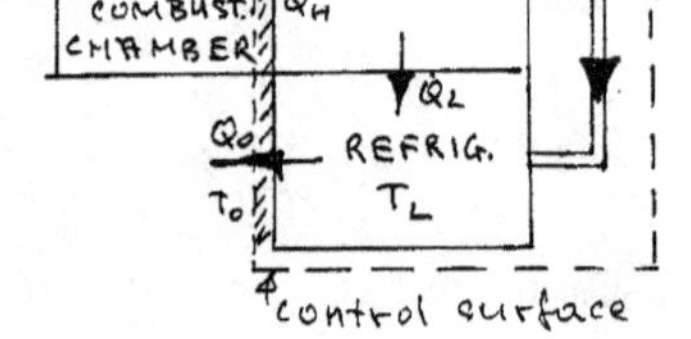

Fig. 3.17

Analysis: Maximum thermal input

$$\left[\dot{Q}_{\text{H}}\right]_{\text{MAX}} = -\dot{m}_{\text{f}} \Delta h_{\text{o}}$$
$$= 0.025 \frac{\text{kg}}{\text{s}} \; 40000 \frac{\text{kJ}}{\text{kg}} = 1000 \text{ kW}$$

The power system defined by the control surface shown above operates between the heat source of T_H and the heat sink, the environment, at T_o. Therefore, the maximum power that this system can deliver is

$$\left[\dot{W}_{NET}\right]_{MAX} = \dot{Q}_H \frac{T_H - T_o}{T_H}$$

$$= 1000 \text{ kW} \frac{1000 - 700}{1000} = \underline{700 \text{ kW}}$$

As this is less than the specified power output of 720 kW, the inventor's claims are unjustified.

Problem 3.18

Process: Production of two steady streams of air, one hot and the other cold, from the stream of dry atmospheric air, all three at atmospheric pressure, in a device which uses only shaft power input.

To determine: The minimum shaft power input and the mass flow rates of the two emerging streams.

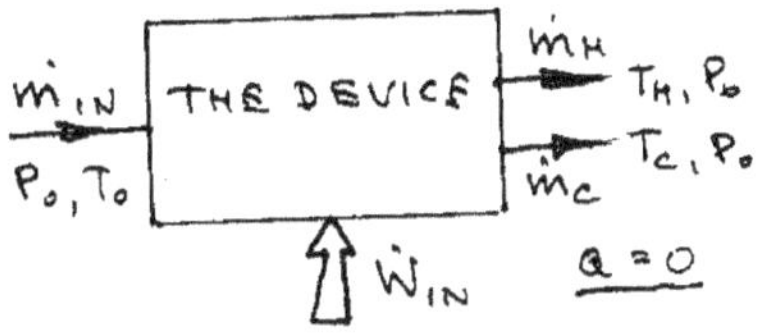

Given data: $\dot{m}_{IN} = 0.1 \text{ kg/s}$ $\quad P_o = 1 \text{ atm}$

$T_o = 25\ °\text{C}$ $\quad T_H = 75\ °\text{C}$

$T_c = 0\ °\text{C}$

Fig.3.18

Assumptions: The device is adiabatic, air is a perfect gas, $\Delta KE = 0$, $\Delta PE = 0$

Analysis: Mass balance

$$\dot{m}_H + \dot{m}_c - \dot{m}_{IN} = 0 \tag{a}$$

Energy equation (assume h_{IN} - 0)

$$\dot{m}_H h_H + \dot{m}_c h_c = \dot{W}_{IN} \tag{b}$$

Exergy balance with $\dot{I} = 0$ (for $\dot{W}_{IN} = (\dot{W}_{IN})_{MIN}$)

and $\quad \varepsilon_{IN} = 0$

$$\dot{m}_H \varepsilon_H + \dot{m}_c \varepsilon_c = W_{IN} \tag{c}$$

Solving (a), (b) and (c) for W_{IN}, $\dot{m}_c$ and $\dot{m}_H$,

$$\dot{W}_{IN} = \dot{m}_{IN} \frac{\varepsilon_H h_c - h_H \varepsilon_c}{T_o (s_c - s_H)} \tag{d}$$

$$\dot{m}_c = \frac{\dot{W}_{IN} - \dot{m}_{IN} h_H}{h_c - h_H} \tag{e}$$

$$\dot{m}_H = \dot{m}_{IN} - \dot{m}_c \qquad \text{(f)}$$

For air as a perfect gas

$$h - h_o = c_P(T - T_o) \qquad \text{(g)}$$

Also, with $P = P_o$ = const

$$s - s_o = c_P \ln T/T_o \qquad \text{(h)}$$

$$\varepsilon^{\Delta T} = \varepsilon = c_P(T - T_o - T_o \ln\frac{T}{T_o}) \qquad \text{(i)}$$

Substituting numerical data

$$\dot{W}_{IN} = 0.2086 \text{ kW}$$

$$\dot{m}_c = 0.064 \text{ kg/s}$$

$$\dot{m}_H = \underline{0.036 \text{ kg/s}}$$

<u>Comment</u>: The solution evaluates the minimum power input and the mass flow rates $\dot{m}_c$ and $\dot{m}_H$ under reversible operating conditions. It also demonstrates that the process is thermodynamically feasible. What it does not do is to show what ideal processes should be used to achieve the required objectives. This is left to the reader's ingenuity and inventiveness.

<u>Problem 3.19</u>

<u>Process</u>: Feed water heating in a closed-type, regenerative heater.

<u>To determine</u>: (i) The bled steam fraction, α

(ii) irreversibility per mass of the steam entering the turbine.

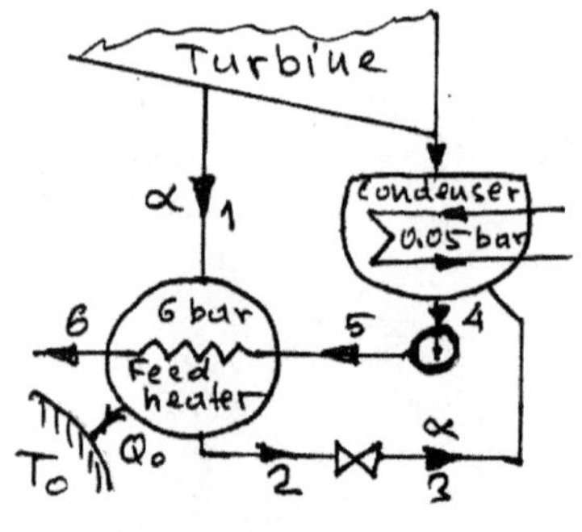

Fig.3.19.a

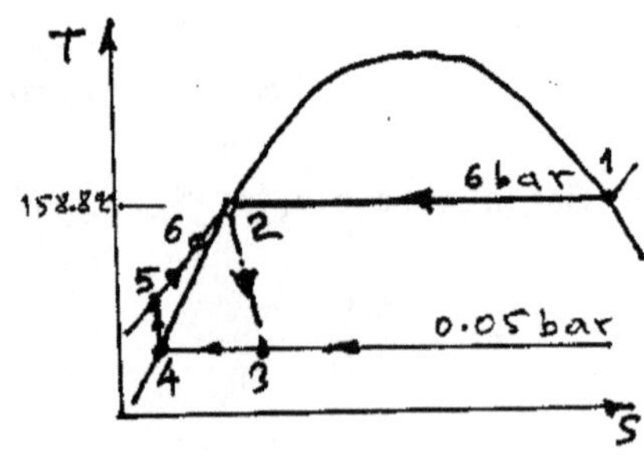

Fig.3.19.b

Given data: $T_2 - T_6 = 7$ K $T_o = 290$ K

$$Q_o = 0.03\ \alpha(h_1 - h_2) \quad \text{(a)}$$

Assumptions: Pressure losses, feed-pump work, $\Delta KE, \Delta PE$ are negligible.

Analysis: Energy balance for the feed heater

$$-Q_o = (\alpha h_2 + h_6) - (\alpha h_1 + h_5) \quad \text{(b)}$$

Substituting (a) in (b) and re-arranging

$$\alpha = \frac{h_6 - h_5}{0.97(h_1 - h_2)} \quad \text{(c)}$$

Irreversibility for the feed heater

$$I = T_o\left[(s_6 - s_5) - \alpha(s_2 - s_2) + \frac{Q_o}{T_o}\right] \quad \text{(d)}$$

Steam properties:

$h_1 = 2757$ kJ/kg $s_1 = 6.761$ kJ/kgK
$h_2 = 670$ kJ/kg $s_2 = 1.931$ kJ/kgK

$h_5 \cong h_4 = 138$ kJ/kg $s_5 \cong s_4 = 0.476$ kJ/kgK

$h_6 = 640$ kJ/kg $s_6 = 1860$ kJ/kgK

Substituting numerical data in (c), (b) and (d)

$$\alpha = \frac{640 - 138}{0.97(2757 - 670)} = \underline{0.248}$$

$$Q_o = 0.03 \times 0.2480\,(2757 - 670) = 15.53 \text{ kJ/kg}$$

$$\dot{I} = 290[(1.860 - 0.476) - 0.248(6.761 - 1.931) + 15.53\}$$

$$= \underline{69.5 \text{ kJ/kg}}$$

Problem 3.20

The process: Heat exchange between a condensing fluid and a boiling fluid.

To derive: $\dot{I} = \dot{Q}\dfrac{T_o \Delta T}{T_M^2}$ where $\Delta T = T_1 - T_2$

$T_M = \sqrt{T_1 T_2}$

Assumptions: t_1 = const. and t_2 = const. and pressure losses are negligible.

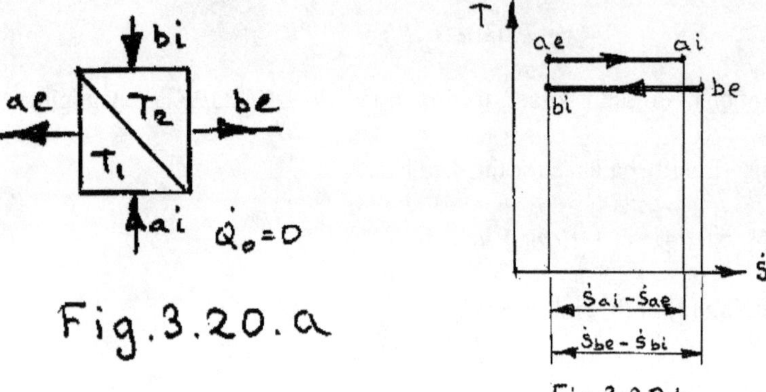

Fig.3.20.a

Fig.3.20.b

Analysis: When $\dot{Q}_\mathrm{i} = 0$

$$\dot{I} = T_\mathrm{o}[(\dot{S}_\mathrm{be} - \dot{S}_\mathrm{bi}) - (\dot{S}_\mathrm{ai} - \dot{S}_\mathrm{ae})] \qquad \text{(a)}$$

Heat transfer rate ($\Delta P = 0$)

$$\dot{Q} = T_1(\dot{S}_\mathrm{ai} - \dot{S}_\mathrm{ae}) = T_2 - (\dot{S}_\mathrm{be} - \dot{S}_\mathrm{bi})$$

$$\therefore \quad \dot{S}_\mathrm{ai} - \dot{S}_\mathrm{ae} = \frac{\dot{Q}}{T_1} \qquad \text{(b)}$$

$$\therefore \quad \dot{S}_\mathrm{be} - \dot{S}_\mathrm{bi} = \frac{\dot{Q}}{T_2} \qquad \text{(c)}$$

Substituting (b) and (c) in (a)

$$\dot{I} = \dot{Q}\frac{T_\mathrm{o}\Delta T}{T_\mathrm{M}^2} \quad \text{where } \Delta T = T_1 - T_2 \text{ and } T_\mathrm{M} = \sqrt{T_1 T_2}$$

Q.E.D.

Problem 4.1

The process: Adiabatic throttling of NH_3 and R-12 between the same temperature limits.

To determine: (i) Specific irreversibilities for the two refrigerants,

(ii) irreversibility rates for the two refrigerants when the refrigeration duty is 100 kW.

Given data: T_o = 300 K $\qquad \dot{Q}_\mathrm{REF} = 100\,\mathrm{kW}$

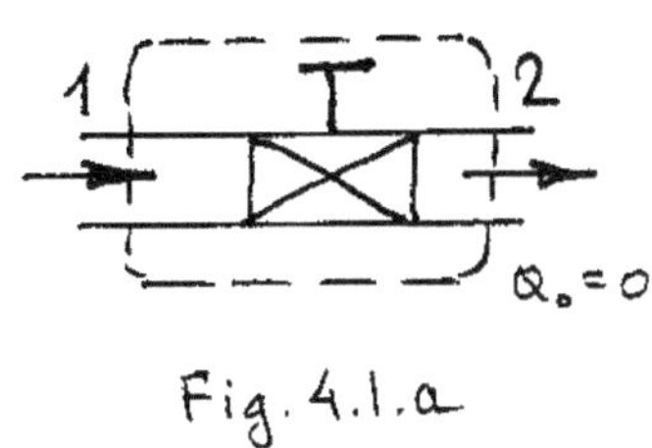

Fig. 4.1.a

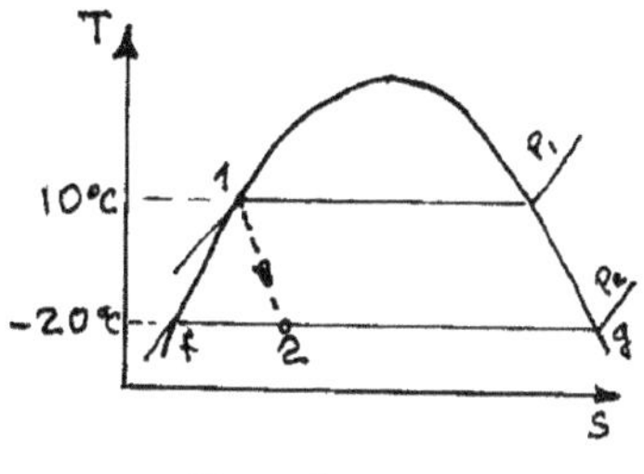

Fig. 4.1.b

Thermodynamic Properties

	Initial State $T_{SAT} = 10\,°C$		Final State $T_{SAT} = -20\,°C$			
	$\frac{h_f}{kJ/kg}$	$\frac{s_f}{kJ/kgK}$	$\frac{h_f}{kJ/kg}$	$\frac{s_f}{kJ/kgK}$	$\frac{h_g}{kJ/kg}$	$\frac{s_g}{kJ/kgK}$
NH_3	227.8	0.881	89.8	0.368	1420.0	5.623
R-12	45.37	0.1752	17.82	0.0731	178.73	0.7087

Assumptions: Pressure losses in the heat transfer process in the evaporator are negligible, the throttling process is adiabatic.

Analysis: From the Gouy-Stodola relation $(Q_o = 0)$

$$\dot{I} = \dot{m}T_o(s_2 - s_1) \qquad \text{(a)}$$

For adiabatic throttling

$$h_1 = h_2 = h_f + x_2(h_g - h_f)$$

$$\therefore \quad x_2 = \frac{h_1 - h_f}{h_g - h_f} \qquad \text{(b)}$$

and

$$s_2 = s_f + x_2\,(s_g - s) \qquad \text{(c)}$$

Specific Irreversibilities

NH_3 substituting in (b) and (c)

$$s_2 = 0.913 \text{ kJ/kgK}$$

From (a) $\quad i = \dot{I}/\dot{m} = 300(0.913 - 0.881)$

$\quad = \underline{9.65 \text{ kJ/kg}}$

R-12 $\quad s_2 = 0.1818$ kJ/kgK

And from (a) $i = \dot{I}/\dot{m} = 300(0.1818 - 0.1752)$

$= \underline{1.976 \text{ kJ/kg}}$

<u>Irreversibility rates for 100 kW refrigeration duty</u>

For the evaporator $\dot{Q}_{REF} = \dot{m}(h_g - h_2) = \dot{m}(h_g - h_1)$

$$\therefore \dot{m} = \frac{\dot{Q}_{REF}}{h_g - h_1}$$

<u>NH_3</u> $\dot{m} = \dfrac{100}{1420 - 227.8} = 0.0839 \text{ kg/s}$

$$\therefore \dot{I} = \dot{m}i = 0.0839 \text{ (kg/s)} \times 9.65 \text{ (kJ/kg)}$$

$= \underline{0.8094 \text{ kW}}$

<u>R-12</u> $\dot{m} = \dfrac{100}{178.73 - 45.37} = 0.7498 \text{ kg/s}$

$$\therefore \dot{I} = \dot{m}i = 0.7498 \text{ (kg/s)} \times 1.976 \text{ (kJ/kg)}$$

$= \underline{1.482 \text{ kW}}$

<u>Comment</u>: The above solution demonstrates the fact that:
(i) internal irreversibility of a process depends on the type of working fluid used,

(ii) a fair comparison can only be made on the basis of the same plant output. Note, however, that although NH_3 has an advantage over R-12 on the basis of this comparison, the situation may be reversed when e.g. the compression process is considered.

<u>Problem 4.2</u>

<u>The process</u>: Expansion of steam in an adiabatic nozzle.

<u>To determine</u>: (i) Steam exit velocity,

(ii) specific irreversibility of the process,

(iii) the rational efficiency.

Given data: $P_1 = 8$ bar $T_1 = 300$ °C

$P_2 = 2$ bar $T_2 = 150$ °C

$T_o = 293$ K

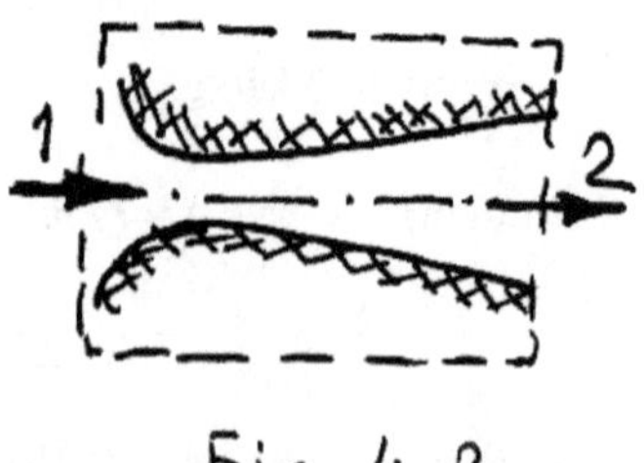

Fig. 4.2

Assumptions: Steam inlet velocity is negligible, the nozzle is adiabatic.

Analysis:

(i) SFEE with $C_1 = 0,\ \dot{Q} = 0,\ \dot{W}_x = 0$

$$0 = \left(h_2 + \frac{C_2^2}{2}\right) - h_1$$

$$\therefore\quad C_2 = \sqrt{2(h_1 - h_2)} \qquad \text{(a)}$$

Properties $h_1 = 3057$ kJ/kg $h_2 = 2770$ kJ/kg

Substituting in (a)

$$C_2 = \sqrt{2 \times 10^3 (3057 - 2770)}$$

$$= \underline{757.6 \text{ m/s}}$$

(ii) Specific entropy production calculated in Problem 1.10

$$\pi = 0.047 \text{ kJ/kgK}$$

Using the Gouy-Stodola relation

$$i = T_o \pi = 293 \times 0.047 = \underline{13.77 \text{ kJ/kg}}$$

(iii) Specific exergy balance for the nozzle

$$\varepsilon_1 - \varepsilon_2 = \frac{C_2^2}{2} + i \qquad \text{(b)}$$

As follows (b) and the function of the nozzle

$$\text{exergy output} = \frac{C_2^2}{2}$$

$$\text{exergy input} = \varepsilon_1 - \varepsilon_2$$

Hence, the rational efficiency can be expressed as

$$\psi = \frac{C_2^2/2}{\varepsilon_1 - \varepsilon_2} = \frac{h_1 - h_2}{h_1 - h_2 - T_o(s_1 - s_2)}$$

$$= \underline{0.954}$$

Problem 4.3

The process: Expansion of air in a turbine.

To determine: (i) Specific irreversibility of the process,

(ii) isentropic efficiency and,

(iii) rational efficiency of the process.

Given data: $P_1 = 3$ bar $P_2 = 1$ bar

$T_1 = 425$ K $T_2 = 340$ K

$T_o = 293$ K

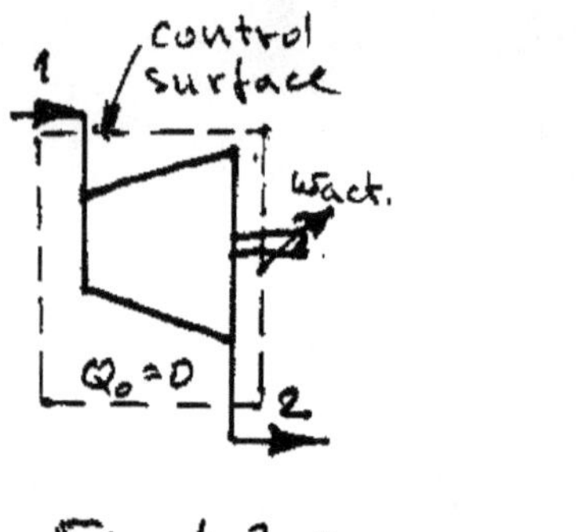

Fig.4.3.a

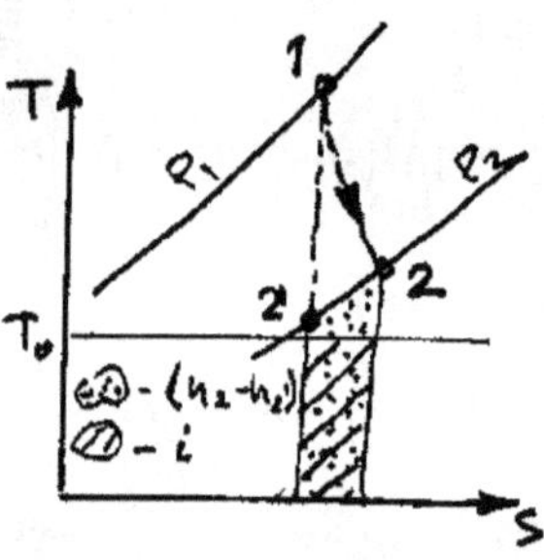

Fig.4.3.b

Assumptions: The turbine is adiabatic, air is a perfect gas, $\Delta KE = 0$, $\Delta PE = 0$

Analysis:

(i) From the Gouy-Stodola relation with $Q = 0$

$$i = \dot{I} / \dot{m} = T_o (s_2 - s_1) \tag{a}$$

For a perfect gas

$$i = T_o \left(c_P \ln \frac{T_2}{T_1} - R \ln \frac{P_2}{P_1} \right) = \underline{26.68 \text{ kJ/kg}}$$

(ii) Isentropic efficiency

$$\eta_s = \frac{W_{\text{ACTUAL}}}{W_{\text{ISENTROPIC}}} = \frac{h_1 - h_2}{h_1 - h_2'} \tag{b}$$

where $W_{\text{ACTUAL}} = c_P (T_1 - T_2)$ (c)

$W_{\text{ISENTROPIC}} = c_P (T_1 - T_2')$ (d)

Also $T_2' = T_1 \left(\frac{P_2}{P_1}\right)^{\frac{\gamma-1}{\gamma}} = 310.5\,\text{K}$

Substituting numerical values in (d), (c) and (b)

$$w_{\text{ACT.}} = 85.4\,\text{kJ/kg}, \quad w_{\text{ISEN.}} = 115.1\,\text{kJ/kg}$$

and

$$\eta_s = \frac{85.4}{115.4} = \underline{0.742}$$

(iii)Rational efficiency

From the exergy balance for the turbine (see the control surface, above)

$$\underbrace{\varepsilon_1 - \varepsilon_2}_{\text{INPUT}} = \underbrace{w_{\text{ACT}} + i}_{\text{OUTPUT}} \tag{e}$$

$$\therefore \quad \frac{\text{OUTPUT}}{\text{INPUT}} = \frac{w_{\text{ACT}}}{\varepsilon_1 - \varepsilon_2} = \frac{w_{\text{ACT}}}{W_{\text{ACT}} + i} \tag{f}$$

Also,

$$\psi = \frac{h_1 - h_2}{(h_1 - h_2) + T_o(s_2 - s_1)} \tag{g}$$

Using (f)

$$\psi = \frac{85.4}{85.4 + 26.68} = \underline{0.762}$$

<u>Comment</u>: Expression (b) can also be written as

$$\eta_s = \frac{h_1 - h_2}{(h_1 - h_2) + (h_2 - h_2')} \tag{h}$$

As follows from (g) and (h), since $(h_1 - h_2') > T_o(s_2 - s_1)$ (see T-S diagram above), therefore, for this expander $\psi > \eta_s$.

Problem 4.4

The process: Control of the power output of a steam turbine by throttling.

To determine:

(i) At full supply pressure:

(a) specific irreversibility of the process
(b) specific work output
(c) rational efficiency

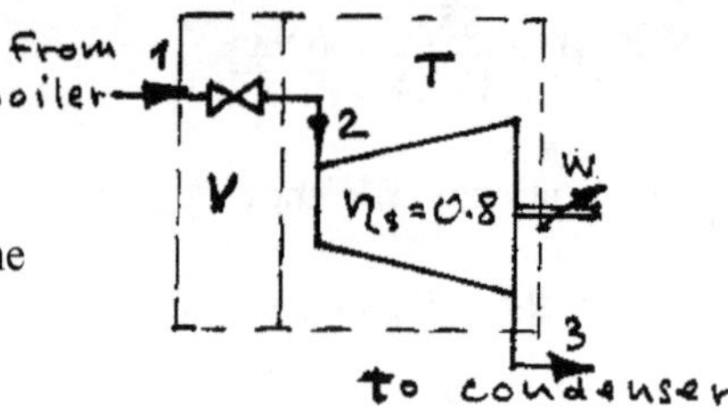

Fig. 4.4.a

(ii) At reduced pressure to 10 bar

(a) sp. irreversibility of valve and turbine
(b) sp. work,
(c) rational efficiency

Given data: $P_1 = 15$ bar $T_1 = 350$ °C

$P_3 = 0.05$ bar $T_o = 290$ K

Assumptions: Both the valve and the turbine operate adiabatically, $\Delta KE = 0$, $\Delta PE = 0$

Analysis:

(i) Operation at full pressure ($i_v = 0$)

From Gouy-Stodola relation for sub region T or V ($Q = 0$)

$$i_T = T_o (s_3 - s_1)$$

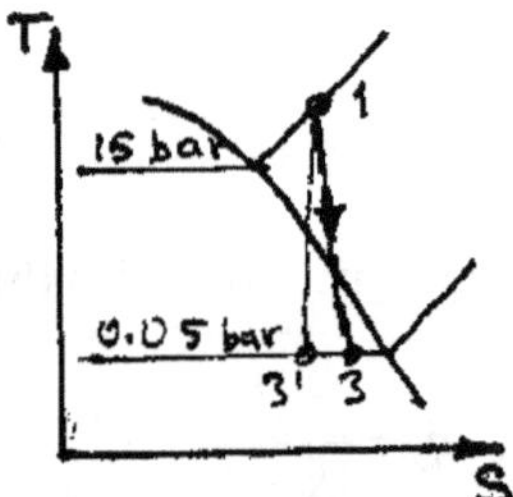

Fig. 4.4.b

From the energy equation

$$w = h_1 - h_3$$

Specific exergy balance $\varepsilon_1 - \varepsilon_3 = w + i$

But, turbine output = w

Hence, turbine input = $\varepsilon_1 - \varepsilon_3$

$$\therefore \psi = \frac{w}{\varepsilon_1 - \varepsilon_3} = \frac{w}{w + i} \qquad \text{(c)}$$

Steam properties:

$h_1 = 3148$ kJ/kg $s_1 = 7.102$ kJ/kgK

At $P_3 = 0.05$ bar: $h_f = 138$ kJ/kg; $h_{fg} = 2423$ kJ/kg

$s_{fg} = 7.918$ kJ/kgK; $s_f = 0.476$ kJ/kgK

$$s_1 = s_{3'} = s_f + x_{3'} s_{fg} \quad \therefore x_{3'} = \frac{7.102 - 0.476}{7.918} = 0.839$$

$$h_{3'} = 138 + 0.839 \times 2423 = 2171 \text{ kJ/kg}$$

$$h_3 = 3148 - 0.8\,(3148-2171) = 2366 \text{ kJ/kg}$$

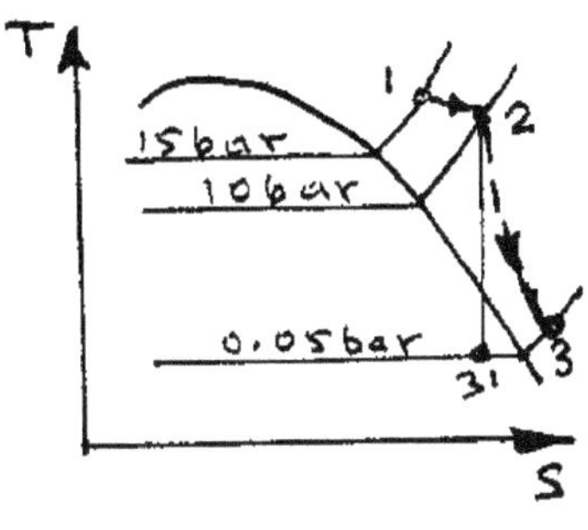

Fig.4.4.c

$$x_3 = \frac{h_3 - h_f}{h_{fg}} = \frac{2366 - 138}{2423} = 0.92$$

$$s_3 = 0.476 + 0.92 \times 7.918 = 7.757 \text{ kJ/kgK}$$

Substituting numerical values in (a), (b) and (c)

(ia) $i_T = 290\,(7.757 - 7.102) = \underline{189.95 \text{ kJ/kg}}$

(ib) $w = 3148 - 2366 = \underline{782 \text{ kJ/kg}}$

(ic) $\psi_T = \dfrac{782}{782 + 189.95} = \underline{0.8045}$

(ii) Operation with reduced inlet pressure

Steam properties

$h_1 = h_2 = 3148$ kJ/kg

$s_2 = 7.284$ kJ/kgK (by interpolation)

$$s_2 = s_3 = s_f + x_{3'}\, s_{fg}$$

$$\therefore \quad x_{3'} = \frac{7.284 - 0.476}{7.918} = 0.86$$

$$h_{3'} = 138 + 0.86 \times 2423 = 2221 \text{ k/kg}$$

$$h_3 = 3148 - 0.8(3148 - 2221) = 2406.4 \text{ kJ/kg}$$

$$x_3 = \frac{2406.4 - 138}{2423} = 0.936$$

$$s_3 = 0.476 + 0.936 \times 7.918 = 7.889 \text{ kJ/kgK}$$

Substituting numerical values in (a), (b) and (c)

(iia) $i_V = 290(7.284 - 7.102) = \underline{52.8 \text{ kJ/kg}}$

$i_T = 290(7.889 - 7.284) = \underline{175.45 \text{ kJ/kg}}$

(iib) $W = 3148\text{-}2406.4 = \underline{741.6 \text{ kJ/kg}}$

(iic) $\psi_T = \dfrac{W}{w + i_T} = \dfrac{741.6}{741.6 + 175.45} = \underline{0.809}$

Alternatively, for the combined system valve and turbine

$$\psi_{\text{V+T}} = \frac{W}{w + i_{\text{T}} + i_{\text{V}}} = \underline{0.765}$$

Comment: Throttling also reduces the mass flow rate of steam which affects directly the shaft power output. This form of power output control is clearly very wasteful as can be seen by comparing ψ_{T} and $\psi_{\text{V+T}}$.

Problem 4.5

The process: Adiabatic throttling of air at cryogenic temperatures.

To determine: Specific irreversibility of throttling processes between two given pressure limits at two different initial temperatures using a ε - h chart (Fig. E2).

Given data: $P_1 = 8$ MPa $P_2 = 0.1$ MPa

Case (a): $T_1 = 140$ K

Case (b): $T_1 = 120$ K

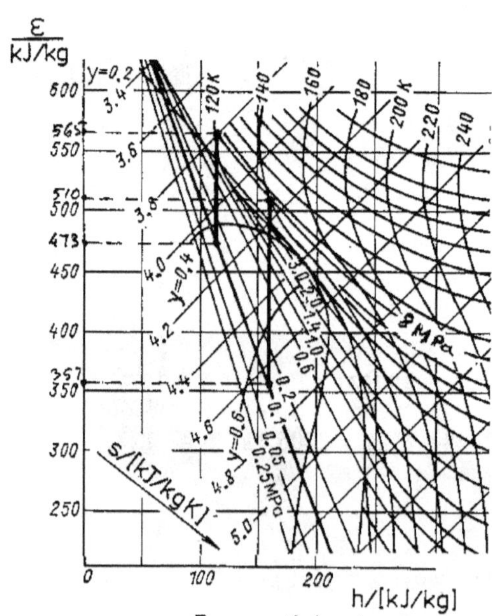

Exergy–enthalpy chart for air.

Fig. 4.5

Analysis:

Since $q = 0$ and $w = 0$, $i = \varepsilon_1 - \varepsilon_2$

Case (a): $\varepsilon_1 = 510$ kJ/kg $\varepsilon_2 = 357$ kJ/kg

$\therefore$ $i = 510 - 357 = \underline{153 \text{ kJ/kg}}$

Case (b): $\varepsilon_1 = 565$ kJ/kg $\varepsilon_2 = 473$ kJ/kg

$\therefore$ $i = 565 - 473 = \underline{92 \text{ kJ/kg}}$

Comment: The lower is the initial temperature of the throttling process, the lower is its irreversibility and the higher is the fraction of the liquid at the end of the throttling process. The effect of this is an increase of the plant rational efficiency.

Problem 4.6

The process: A steam operated ejector.

To determine: (i) The state of the steam leaving the mixing chamber,
(ii) irreversibility rate of the mixing process.

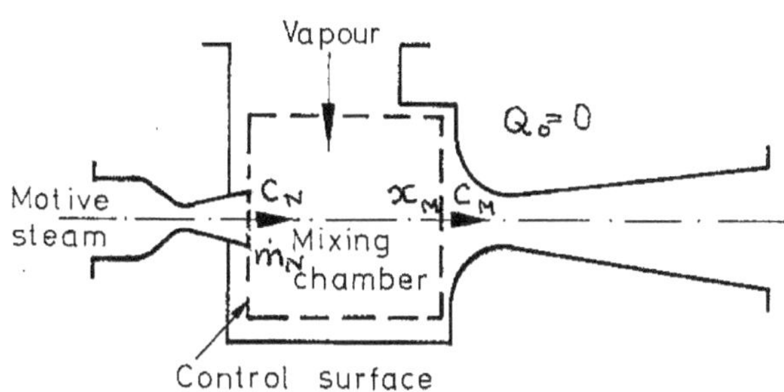

Fig. 4.6

Given data:

$\dot{m}_N = 1.2$ kg/s $\qquad C_N = 1100$ m/s

$P_N = 0.091$ bar $\qquad x_N = 0.85$

$\dot{m}_V = 0.6$ kg/s $\qquad C_V = 150$ m/s

$T_o = 293$ K

Assumptions: Vapour enters the mixing chamber at right angle to the nozzle axis, mixing is adiabatic at constant pressure and with conservation of axial momentum.

Analysis:

For the control surface shown, equations of

continuity $\quad \dot{m}_M = \dot{m}_V + \dot{m}_N \qquad$ (a)

momentum $\quad \dot{m}_N C_N = (\dot{m}_V + \dot{m}_N) C_M \qquad$ (b)

energy $\quad \dot{m}_V (h_V + C_V^2/2) + \dot{m}_N (h_N + C_N^2/2) = \dot{m}_M (h_M + C_M^2/2) \qquad$ (c)

Properties: At 0.091 bar

$h_f = 184.2$ kJ/kg; $\quad h_{fg} = 2397$ kJ/kg; $\quad h_g = 2581$ kJ/kg;

$s_f = 0.625$ kJ/kgK; $\quad s_{fg} = 7.558$ kJ/kgK; $\quad s_g = 8.183$ kJ/kgK

$\therefore \; h_{N'} = 184.2 + 0.85 \times 2397 = 2221.7$ kJ/kg

$s_N = 0.625 + 0.85 \times 7.558 = 7.049$ kJ/kgK

$s_V = s_g = 8.183$ kJ/kgK; $h_V = h_g = 2581$ kJ/kg

From (b) $$C_M = \frac{1.2 \times 1100}{(0.6 + 1.2)} = \underline{733.3 \text{ m/s}}$$

From (c) $$h_M = \frac{0.6}{1.8}(2581 + \frac{150^2}{2} \times 10^{-3}) + \frac{1.2}{1.8}\left(2221.7 + \frac{1100^2}{2} \times 10^{-3}\right) - \frac{733.3^2}{2 \times 10^3}$$

$$= 2479.57 \text{ kJ/kg}$$

Dryness fraction of the mixture

$$x_M = \frac{h_M - h_f}{h_{fg}} = \underline{0.96}$$

$$\therefore \quad s_M = s_f + x_M s_{fg} = 7.5625 \text{ kJ/kgK}$$

Irreversibility rate, from the Gouy-Stodola relation ($\dot{Q} = 0$)

$$\dot{I} = T_o[\dot{m}_M s_M - (\dot{m}_V s_V + \dot{m}_N s_N)]$$

$$= 293[1.8 \times 7.8625 - 0.6 \times 8.183 - 1.2 \times 7.0493]$$

$$= \underline{229.6 \text{ kW}}$$

Comment: The calculated irreversibility rate represents only part of the total irreversibility rate which takes place inside the ejector. Another important contribution to total irreversibility takes place in the diffuser, where some of the kinetic energy of the mixture is recovered.

Problem 4.7

The process: Two-stage adiabatic compression without intercooling.

To determine: (i) the irreversibility rate,

(ii) net power input, and

(iii)the rational efficiency, for each compressor.

Given data: $\dot{m} = 0.3$ kg/s $\quad P_i = P_o = 1$ bar

$T_i = T_o = 300$ K $\quad P_2 = 3$ bar

$P_3 = 9$ bar $\quad \eta_{1,2} = \eta_{2,3} = 0.75$

Assumptions: The compressors are adiabatic, air behaves as a perfect gas, $\Delta KE = 0,\ \Delta PE = 0$

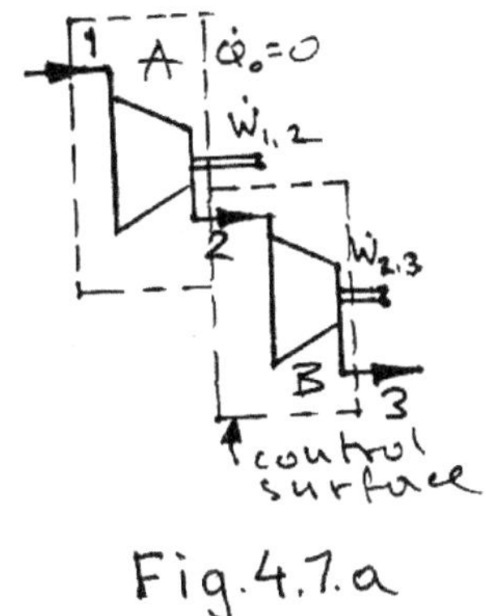

Fig.4.7.a

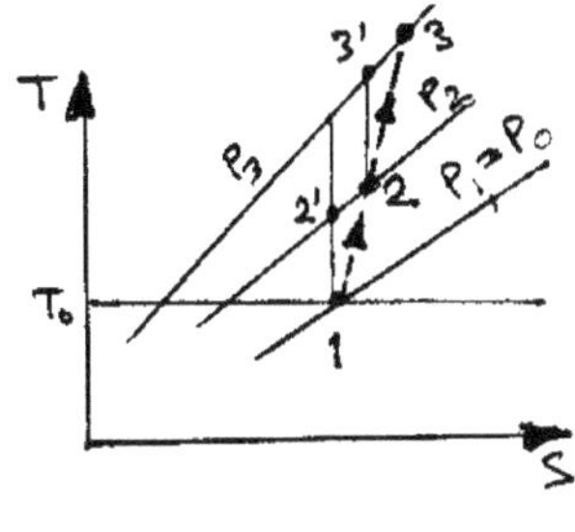

Fig.4.7.b

Analysis:

Irreversibility rate: from G-S relation with $\dot{Q}=0$

$$\dot{I}=\dot{m}T_o(s_e-s_i) \tag{a}$$

where, for a perfect gas

$$s_e-s_i=c_P\ln\frac{T_e}{T_i}-R\ln\frac{P_e}{P_i} \tag{b}$$

Power input

$$\dot{W}_{i,e}=\dot{m}c_P(T_R-T_i) \tag{c}$$

Rational efficiency, from the exergy balance

$$\underbrace{\dot{E}_e-\dot{E}_i}_{\text{OUTPUT}}=\underbrace{\dot{W}_c}_{\text{INPUT}}-\dot{I} \tag{d}$$

$$\psi=\frac{\dot{E}_e-\dot{E}_i}{\dot{W}_c}=1-\frac{\dot{I}}{\dot{W}_c} \tag{e}$$

For an isentropic compression, 1st stage

$$T_{2'}=T_1\left(\frac{P_2}{P_1}\right)^{\frac{\gamma-1}{\gamma}}=410.6\text{ K} \tag{f}$$

From an expression for isentropic efficiency

$$T_2=T_1+\frac{1}{\eta_s}(T_{2'}-T_1) \tag{g}$$

Substituting

$T_2=447.5$ K

Similarly for the 2nd stage

$T_{3'} = 612.5$ K and $T_3 = 667.5$ K

Using (a), (b), (f) and (g) it can be shown that for either stage

$$\dot{I} = \dot{m} T_o c_P \ln\left\{\left[1 + \frac{1}{\eta_s}\left(R_P^{\frac{\gamma-1}{\gamma}} - 1\right)\right] R_P^{\frac{\gamma-1}{\gamma}}\right\} \quad \text{(h)}$$

where

$$R_P = \frac{P_2}{P_1} = \frac{P_3}{P_2} = 3$$

Substituting

$$\underline{\dot{I}_A = \dot{I}_B = 7.781\,\text{kW}}$$

Power input

From (c) $\dot{W}_{1,2} = 0.3 \times 1.005 \times 147.5 = \underline{44.47\text{ kW}}$

$\dot{W}_{2,3} = 0.3 \times 1.005 \times 220 = \underline{66.33\text{ kW}}$

Rational efficiencies

Using the second form of (e)

$$\psi_A = 1 - \frac{7.781}{44.47} = \underline{0.825}$$

$$\psi_B = 1 - \frac{7.781}{66.33} = \underline{0.883}$$

Problem 4.8

The process: Two-stage adiabatic compression of air, with intercooling.

To determine: The effect of the introduction of perfect intercooling on the performance of the plant described in Problem 4.7.

Given data: Plant data as in Problem 4.7 with added perfect intercooling.

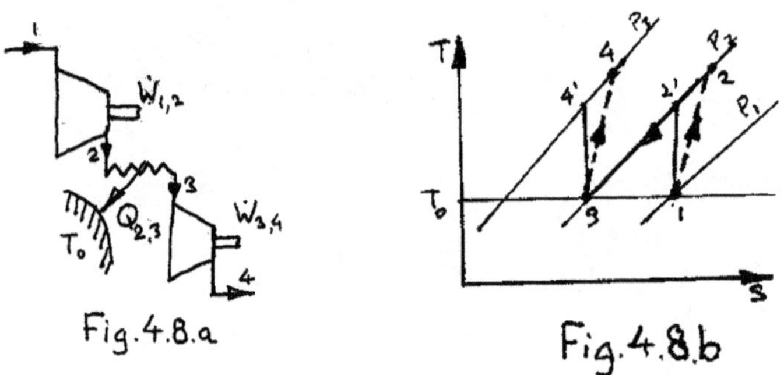

Fig.4.8.a Fig.4.8.b

Analysis:

1st stage: The first stage results will be unaffected by the addition of intercooling.

2nd stage: Because of the perfect gas assumption, the results for the second stage will be the same as those obtained for the first stage.

Problem 4.9

The process: Expansion of steam in an adiabatic turbine.

To derive: An expression for irreversibility rate in the form

$$\dot{I} = f(\dot{W}, \eta_s, T_o T_c \qquad \text{(a)}$$

Derivation: From the Gouy-Stodola relation with $Q = 0$.

$$\dot{I} = \dot{m}T_o(s_2 - s_1) \qquad \text{(a)}$$

but $\quad s_{2'} = s_1 \qquad$ (b)

and $\quad h_2 - h_2' = T_c\,(s_2 - s_1) \qquad$ (c)

Hence $\quad \dot{I} = \dot{m}(h_2 - h_{2'})\dfrac{T_o}{T_c} \qquad$ (d)

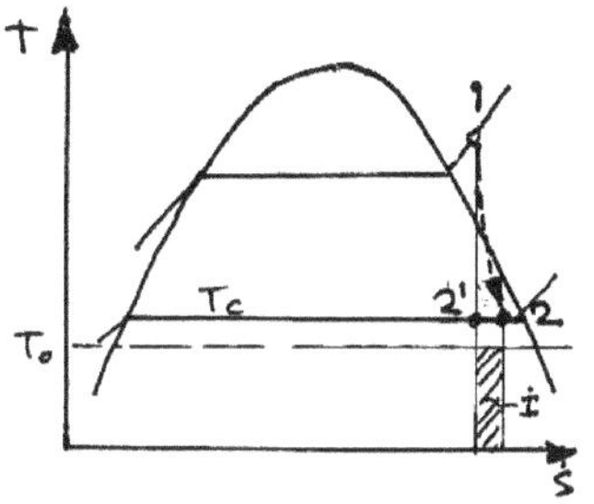

Isentropic efficiency

$$\eta_s = \frac{h_1 - h_2}{h_1 - h_{2'}} = \frac{h_1 - h_2}{(h_1 - h_2)(h_2 - h_{2'})} \qquad \text{(e)}$$

Fig.4.9

From (e)

$$h_2 - h_{2'} = (h_1 - h_2)\left(\frac{1}{\eta_s} - 1\right) \qquad \text{(f)}$$

Substituting (f) in (d)

$$\dot{I} = \dot{m}(h_1 - h_2)\left(\frac{1}{\eta_s} - 1\right)\frac{T_o}{T_c}$$

QED (g)

Also:

Power output of the turbine

Is $\quad \dot{W} = \dot{m}(h_1 - h_2) \qquad$ (h)

$$\dot{I} = \dot{W}\left(\frac{1}{\eta_s} - 1\right)\frac{T_o}{T_c} \qquad \text{(i)}$$

Problem 4.10

The process: Heating of a stream of air by steam from a back-pressure plant, in counter-flow.

To determine: Formulate the rational efficiency and calculate:

(i) mass flow rate of air,

(ii) irreversibility rate, and

(iii) the rational efficiency of the heat transfer process.

Given data: $\dot{m}_w = 0.4$ kg/s State 1w: dry sat.

$P_w = 0.2$ bar $T_{2w} = 41.5$ °C

$P_{1a} = 1.02$ bar $T_{1a} = 10$ °C

$P_{2a} = 0.99$ bar $T_{2a} = 50$ °C

$T_o = 10$ °C

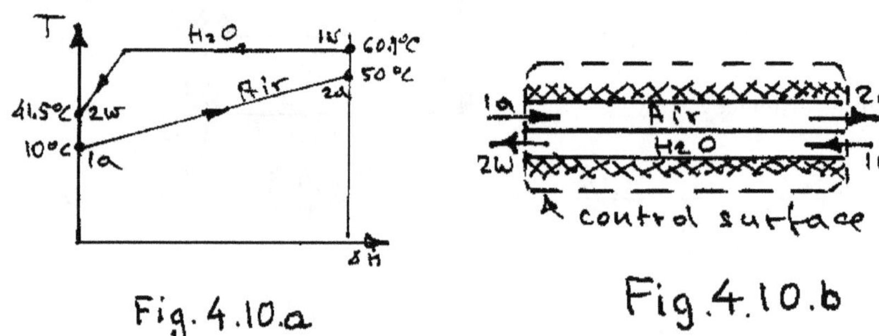

Fig.4.10.a Fig.4.10.b

Assumptions: Air behaves as a perfect gas, $\Delta KE = 0$, $\Delta PE = 0$, pressure losses in the H_2O stream and heat losses from the heat exchanger are negligible.

Analysis:

Derivation of the rational efficiency

Exergy balance for the control region

$$\dot{I} = \left(\dot{E}_{1w} - \dot{E}_{2w}\right) - \left(\dot{E}_{2a} - \dot{E}_{1a}\right) \tag{a}$$

where

$$\dot{E}_{1w} - \dot{E}_{2w} = \dot{m}_w\left[\left(h_{1w} - h_{2w}\right) - T_o\left(s_{1w} - s_{2w}\right)\right]$$

$$= \Delta\dot{E}_w^{\Delta T} \text{ (since } P_w = \text{const)} \tag{b}$$

$$\dot{E}_{2a} - \dot{E}_{1a} = \dot{m}_a c_P\left\{\left[(T_{2a} - T_{1a}) - T_o \ln\frac{T_{2a}}{T_{1a}}\right] - \frac{\gamma - 1}{\gamma} T_o \ln\frac{P_{1a}}{P_{2a}}\right\}$$

$$= \Delta\dot{E}_a^{\Delta T} - \Delta\dot{E}_a^{\Delta P} \tag{c}$$

where $$\Delta \dot{E}_{\mathrm{a}}^{\Delta T} = \dot{m}_a c_P \left[(T_{2\mathrm{a}} - T_{1\mathrm{a}}) - T_{\mathrm{o}} \ln \frac{T_{2\mathrm{a}}}{T_{1\mathrm{a}}} \right] \quad \text{(d)}$$

$$\Delta \dot{E}_{\mathrm{a}}^{\Delta P} = \dot{m}_a c_P \frac{\gamma - 1}{\gamma} T_{\mathrm{o}} \ln \frac{P_{1\mathrm{a}}}{P_{2\mathrm{a}}} \quad \text{(e)}$$

Substituting (b), (c) and (d) in (a)

$$\underbrace{\Delta \dot{E}_{\mathrm{a}}^{\Delta T}}_{\text{OUTPUT}} = \underbrace{\Delta \dot{E}_{\mathrm{w}}^{\Delta T} + \Delta \dot{E}_{\mathrm{a}}^{\Delta P}}_{\text{INPUT}} - \dot{I} \quad \text{(f)}$$

Hence, the rational efficiency

$$\psi = \frac{\text{OUTPUT}}{\text{INPUT}} = \frac{\Delta \dot{E}_{\mathrm{a}}^{\Delta T}}{\Delta \dot{E}_{\mathrm{w}}^{\Delta T} + \Delta \dot{E}_{\mathrm{a}}^{\Delta P}} \quad \text{(g)}$$

or

$$\psi = 1 - \frac{\dot{I}}{\Delta \dot{E}_{\mathrm{w}}^{\Delta T} + \Delta \dot{E}_{\mathrm{a}}^{\Delta P}} \quad \text{(h)}$$

H_2O properties

$h_{1\mathrm{w}} = 2609$ kJ/kg $\qquad s_{1\mathrm{w}} = 7.907$ kJ/kgK

$h_{2\mathrm{w}} = 251$ kJ/kg $\qquad s_{2\mathrm{w}} = 0.832$ kJ/kgK

(i) Mass flow rate (from energy balance)

$$\dot{m}_{\mathrm{a}} = \dot{m}_{\mathrm{w}} \frac{h_{1\mathrm{w}} - h_{2\mathrm{w}}}{c_p (T_{2\mathrm{a}} - T_{1\mathrm{a}})} = 0.4 \frac{2609 - 251}{1.005(50 - 10)} = \underline{23.46 \text{ kg/s}}$$

(ii) Irreversibility rate

$$\Delta \dot{E}_{\mathrm{w}}^{\Delta T} = 0.4[(2609 - 251) - 283.15(7.907 - 0.832)] = 141.9\,\text{kW}$$

$$\Delta \dot{E}_{\mathrm{a}}^{\Delta T} = 23.58 \times 1.005 \left[(50 - 10) - 283.15 \ln \frac{323.15}{283.15} \right] = 60.95\,\text{kW}$$

$$\Delta \dot{E}_{\mathrm{a}}^{\Delta P} = 23.58 \times 1.005 \frac{0.4}{1.4} 283.15 \ln \frac{1.02}{0.99} = 56.95\,\text{kW}$$

Substituting

$$\dot{I} = 141.9 - 60.95 + 56.95 = \underline{137.9\text{W}}$$

(iii) Rational efficiency

From (g) or (h) $\quad \underline{\psi = 0.307}$

Problem 4.11

The process: Chilling a stream of water by evaporating ammonia.

To determine: (i) irreversibility rate of the process,
(ii) the rational efficiency.

Given data: $\dot{m}_w = 1\,\text{kg/s}$ $\dot{m}_a = 0.083\,\text{kg/s}$

Dryness fractions of NH_3

$x_{1a} = 0.18$ $x_{2a} = 1.0$

$P_a = 3.983$ bar $T_o = 300$ K $T_a = 271.15$ K

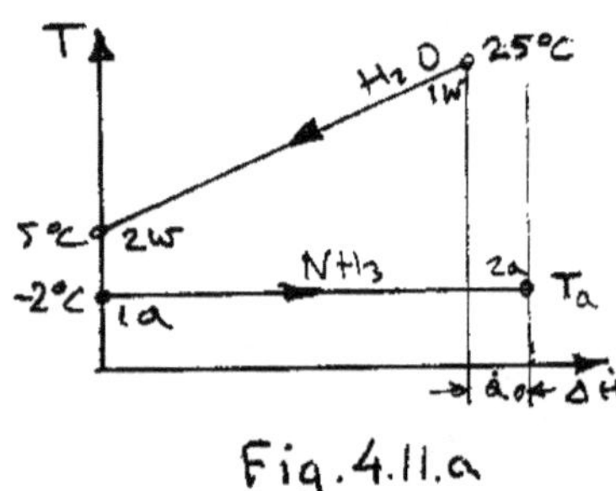

Fig.4.11.a

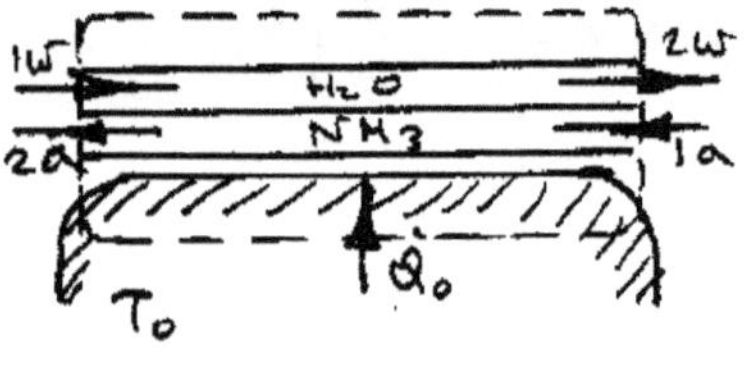

Fig.4.11.b

Assumptions: Pressure losses and ΔKE and ΔPE are negligible.

Analysis: From the Gouy-Stodola relation

$$\dot{I} = T_o\left[\left(\dot{S}_{OUT} - \dot{S}_{IN}\right) - \frac{\dot{Q}_o}{T_o}\right]$$

$$= T_o\left[\dot{m}_a\left(s_{2a} - s_{1a}\right) - \dot{m}_w\left(s_{1w} - s_{2w}\right)\right] - \dot{Q}_o \quad \text{(a)}$$

where $\dot{Q}_o = T_o\left[\dot{m}_a\left(h_{2a} - h_{1a}\right) - \dot{m}_w\left(h_{1w} - h_{2w}\right)\right]$ (b)

In the absence of pressure losses

Exergy input

$$\Delta\dot{E}_{IN} = \dot{m}_a\left(\varepsilon_{1a} - \varepsilon_{2a}\right)$$

$$= \dot{m}_a\left[\left(h_{1a} - h_{2a}\right) - T_o\left(s_{1a} - s_{2a}\right)\right] \quad \text{(c)}$$

$$\therefore \quad \psi = 1 - \frac{\dot{I}}{\Delta\dot{E}_{IN}} \quad \text{(d)}$$

Thermodynamic properties

$h_{1w} = 104.8$ kJ/kg $s_{1w} = 0.367$ kJ/kgK

h_{2w} = 21.0 kJ/kg $\quad$ s_{2w} = 0.076 kJ/kgK

$h_{f,a}$ = 172.0 kJ/kg $\quad$ $s_{f,a}$ = 0.681 kJ/kgK

$h_{g,a} = h_{2a}$ = 1442.2 kJ/kg $\quad$ $s_{g,a} = s_{2,a}$ = 5.365 kJ/kgK

$h_{1a} = (1\text{-}x_{1a})h_{fa} + x_{1a}h_{g,a}$ = 400.8 kJ/kg

$s_{1a} = (1\text{-}x_{1a})\, s_{fa} + x_{1a}s_{g,a}$ = 1.524 kJ/kg

Heat transfer with the environment

From (b)

$$\dot{Q}_o = 0.083(1442.2 - 400.8) - 1.0(104.8 - 21)$$

$$= 2.64 \text{ kW}$$

(i) Irreversibility rate

From (a)

$$\dot{I} = 300[0.083(5.365 - 1.524) - 1.0(0.367 - 0.076)] - 2.64$$

$$= \underline{5.70 \text{ kW}}$$

(ii) The rational efficiency

Exergy input, from (c), with

$$s_{1a} - s_{2a} = \frac{h_{1a} - h_{2a}}{T_a} \qquad \text{(e)}$$

$$\Delta\dot{E}_{IN} = \dot{m}\left[(h_{1a} - h_{2a}) - \frac{T_o}{T_a}(h_{1a} - h_{2a})\right]$$

$$= \dot{m}_a\left(1 - \frac{T_o}{T_a}\right)(h_{1a} - h_{2a}) \qquad \text{(f)}$$

$$\Delta\dot{E}_{IN} = 0.083\left(1 - \frac{300}{271.15}\right)(-1041.4) = 9.197 \text{ kW}$$

From (d) $\quad \Psi = 1 - \dfrac{5.70}{9.197} = \underline{0.380}$

Problem 4.12

The process: Heat transfer from products of combustion to an air stream, in counter-flow.

To determine: (i) Heat transfer rate to the surroundings,

(ii) irreversibility rate of the process and

(iii) the rational efficiency.

Given data:

	c_p/(kJ/kgK)	γ
Air	1.0	1.4
Products of Combustion	1.15	1.333

	Mass flow rate in kg/s	Inlet		Outlet	
		Temp. *T/K*	Press.*P*/bar	Temp. *T/K*	Press.*P*/bar
Products of combustion	0.50	403	1.05	310	1.00
Air	0.55	283	1.10	358	1.05

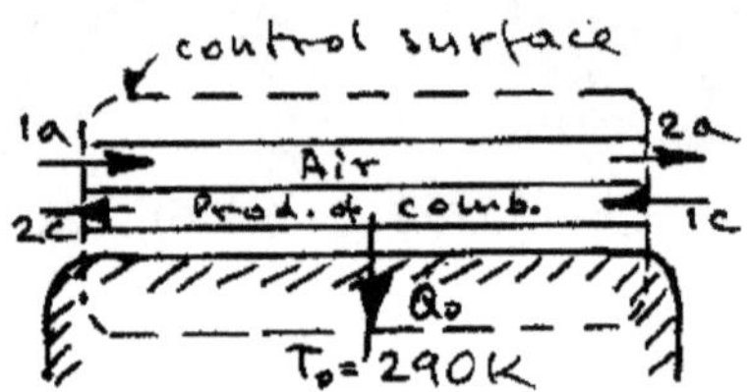

Fig.4.12.a

Assumptions: Both gases approximate to perfect gas behaviour.
Analysis:

(i) Energy balance for the control region

$$\dot{Q} = \dot{m}_c c_{Pc}(T_{2c} - T_{1c}) + \dot{m}_a c_{Pa}(T_{2a} - T_{1a}) \qquad \text{(a)}$$

On substituting numerical data

$$\dot{Q} = \underline{-12.22 \text{ kW}} = -\dot{Q}_o$$

(ii) From the Gouy-Stodola relation

$$\dot{I} = T_o\left[\dot{m}_a(s_{2a} - s_{1a}) + \dot{m}_c(s_{2c} - s_{1c})\right] + \dot{Q}_o \qquad \text{(b)}$$

where, for perfect gases

$$s_2 - s_1 = c_P\left[\ln\frac{T_2}{T_1} - \frac{\gamma - 1}{\gamma}\ln\frac{P_2}{P_1}\right] \qquad \text{(c)}$$

On substituting numerical data in (c) and (b)

$$\dot{I} = \underline{9.136 \text{ kW}}$$

(iii) Rational efficiency

The exergy balance for a heat exchanger involving gas streams can be written in the following form:

$$\dot{I} = \underbrace{\Delta\dot{E}_c^{\Delta T} + \Delta\dot{E}_c^{\Delta P} + \Delta\dot{E}_a^{\Delta P}}_{\text{INPUT}} - \underbrace{\Delta\dot{E}_a^{\Delta T}}_{\text{OUTPUT}} \tag{d}$$

When the streams are perfect gases

$$\Delta\dot{E}_a^{\Delta T} = \dot{m}_a c_{Pa}\left[T_{2a} - T_{1a} - T_o \ln\frac{T_{2a}}{T_{1a}}\right] \tag{e}$$

$$\Delta\dot{E}_c^{\Delta T} = \dot{m}_c c_{Pc}\left[T_{1c} - T_{2c} - T_o \ln\frac{T_{1c}}{T_{2c}}\right] \tag{f}$$

$$\Delta\dot{E}_a^{\Delta P} = \dot{m}_a c_{Pa} T_o \frac{\gamma - 1}{\gamma} \ln\frac{P_{1a}}{P_{2a}} \tag{g}$$

$$\Delta\dot{E}_c^{\Delta P} = \dot{m}_c c_{Pc} T_o \frac{\gamma - 1}{\gamma} \ln\frac{P_{1c}}{P_{2c}} \tag{h}$$

The rational efficiency for the heat exchanger is

$$\Psi = \frac{\text{OUTPUT}}{\text{INPUT}} = \frac{\Delta\dot{E}_a^{\Delta T}}{\Delta\dot{E}_c^{\Delta T} + \Delta\dot{E}_c^{\Delta P} + \Delta\dot{E}_a^{\Delta P}} \tag{i}$$

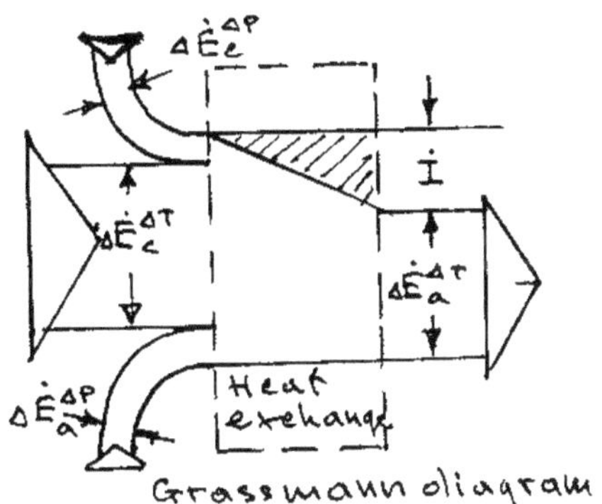

Fig.4.12.b

Substituting numerical data in (e)

$$\text{OUTPUT} = \Delta\dot{E}_a^{\Delta T} = 3.754 \text{ kW}$$

From (d)

$$\text{INPUT} = \Delta\dot{E}_a^{\Delta T} + \dot{I} = 3.754 + 9.136 = 12.89 \text{ kW}$$

$$\therefore \qquad \Psi = \frac{3.754}{12.89} = \underline{0.29}$$

Problem 4.13

The process: Heat exchange as described in Problem 4.12 but with negligible heat loss, pressure losses and near-zero temperature difference at one end.

To determine: (i) Intrinsic irreversibility rate and

(ii) maximum rational efficiency.

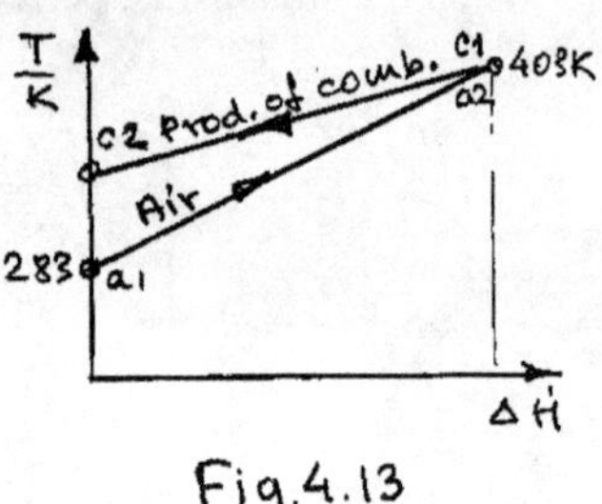

Fig.4.13

Given data: $\dot{m}_{\mathrm{a}} = 0.55\,\mathrm{kg/s}$ $\quad$ $\dot{m}_{\mathrm{c}} = 0.50\,\mathrm{kg/s}$

$c_{P\mathrm{a}} = 1.0$ kJ/kgK $\quad$ $\gamma_{\mathrm{a}} = 1.4$

$c_{P\mathrm{c}} = 1.15$ kJ/kgK $\quad$ $\gamma_{\mathrm{c}} = 1.333$

Assumptions: The heat exchanger is adiabatic, negligible pressure losses, near-zero temperature difference at one end, $\Delta KE = 0$ and $\Delta PE = 0$.

Analysis:

Heat capacity rates

$$\dot{m}_{\mathrm{a}} c_{P\mathrm{a}} = 0.55\,\mathrm{kW/K}, \qquad \dot{m}_{\mathrm{c}} c_{P\mathrm{c}} = 0.575\,\mathrm{kW/K}$$

Hence, since $\dot{m}_{\mathrm{c}} c_{P\mathrm{c}} > \dot{m}_{\mathrm{a}} c_{P\mathrm{a}}$, the minimum temperature difference occurs at the hot end of the heat exchangers, and therefore $T_{\mathrm{c1}} = T_{\mathrm{a2}}$.

From the energy equation

$$T_2 = T_1 - \frac{\dot{m}_a c_{P\mathrm{a}}(T_1 - T_3)}{\dot{m}_c c_{P\mathrm{c}}} = 288.2\,\mathrm{K}$$

(i) Intrinsic irreversibility

From the Gouy-Stodola relation

$$\dot{I} = T_{\mathrm{o}}\left[\dot{m}_{\mathrm{a}}(s_{\mathrm{a2}} - s_{\mathrm{a1}}) + \dot{m}_{\mathrm{c}}(s_{\mathrm{c2}} - s_{\mathrm{c1}})\right] \qquad \text{(a)}$$

For perfect gases, ($P_{\mathrm{a}} = $ const, $P_{\mathrm{c}} = $ const)

$$\dot{I}_{\mathrm{INT}} = T_{\mathrm{o}}\left[\dot{m}_{\mathrm{a}} c_{P\mathrm{a}} \ln\frac{T_{\mathrm{a2}}}{T_{\mathrm{a1}}} + \dot{m}_{\mathrm{c}} c_{P\mathrm{c}} \ln\frac{T_{\mathrm{c2}}}{T_{\mathrm{c1}}}\right] \qquad \text{(b)}$$

$$= 290\left[0.55 \times 1.0 \ln\frac{403}{283} + 0.5 \times 1.15 \ln\frac{288.2}{403}\right]$$

$$= \underline{0.464\,\mathrm{kW}}$$

(ii) Maximum rational efficiency

Since $\dot{I}_{\mathrm{INT}}$ has already been calculated, we can use the following convenient form for Ψ_{MAX}

$$\Psi_{\text{MAX}} = \frac{\Delta \dot{E}_{\text{OUT}}}{\Delta \dot{E}_{\text{OUT}} + \dot{I}_{\text{INT}}} \tag{c}$$

$$\Delta \dot{E}_{\text{OUT}} = \Delta \dot{E}_{\text{a}}^{\Delta T} = \dot{m}_{\text{a}} c_{P\text{a}} \left[(T_{\text{a2}} - T_{\text{a1}}) - T_{\text{o}} \ln \frac{T_{\text{a2}}}{T_{\text{a1}}} \right] \tag{d}$$

$$= 0.55 \times 1.0 \left[(403 - 283) - 290 \ln \frac{403}{283} \right]$$

$$= 9.62 \text{ kW}$$

Hence,

$$\Psi_{\text{MAX}} = \frac{9.62}{9.62 + 0.464} = \underline{0.954}$$

Comment: The reduction in the irreversibility rate, or avoidable irreversibility rate, is

$$\dot{I}_{\text{AVOID}} = \dot{I} - \dot{I}_{\text{INT}}$$

$$= 9.136 - 0.464 = \underline{8.672 \text{ kW}}$$

Such a large reduction is possible because of the very nearly equal values of heat capacity rates.

Problem 4.14

The process: Mixing of heptane and octane.

To determine: (i) Standard molar chemical exergy of the mixture,
(ii) irreversibility of the process per mole of the mixture.

Given data: $T_1 = T_2 = 25\ °\text{C}$ $T_{\text{o}} = 25\ °\text{C}$

$P_{\text{o}} = 1$ atm

	x_i	$\tilde{\varepsilon}_i^{\text{o}}$ /[kJ/kmol]
Heptane, C_7H_{16}	0.9	4786300
Octane C_8H_{18}	0.1	5440030

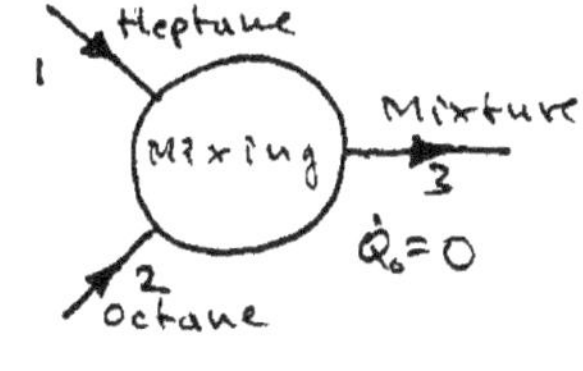

Fig.4.14

Assumptions: The process is adiabatic and involves no change in temperature, i.e. the solution is ideal, $\Delta KE = 0$ and $\Delta PE = 0$.

Analysis:

(i) For an ideal solution

$$\tilde{\varepsilon}_{\text{M}}^{\text{o}} = \sum x_i \tilde{\varepsilon}_i^{\text{o}} + \tilde{R} T^o \sum_i x_i \ln x_i \tag{a}$$

$$= 0.9 \times 4786300 + 0.1 \times 5440030$$

$$+ 8.3144 \times 298.15\ (0.9 \ln 0.9 + 0.1 \ln 0.1)$$

$= 4851673 - 806$

$= \underline{4850867 \text{ kJ/kmol}}$

(ii) Exergy balance for the control region

$$I = x_{\mathrm{H}}\tilde{\varepsilon}_{\mathrm{H}}^{\mathrm{o}} + x_{\mathrm{o}}\tilde{\varepsilon}_{\mathrm{o}}^{\mathrm{o}} - \tilde{\varepsilon}_{\mathrm{M}}^{\mathrm{o}} \qquad \text{(b)}$$

From (a) and b)

$$I = -\tilde{R}T^{\mathrm{o}}[x_{\mathrm{o}} \ln x_{\mathrm{o}} + x_{\mathrm{H}} \ln x_{\mathrm{H}}]$$

$= \underline{806 \text{ kJ/kmol}}$

Problem 4.15

The process: Air separation process into two streams, one of which has the molar composition 90% O_2 and 10% N_2.

To determine: (i) The composition of the other stream,

(ii) minimum exergy input rate to the plant

(iii) also determine (ii) using the ε-x chart for air.

Given data:

$x_{N2} = 0.79 \;\; x_{o2} = 0.21$

$P^1 = P_A = P_B = P_o = 1$ bar

$T^1 = T_A = T_B = T_o = 293$ K

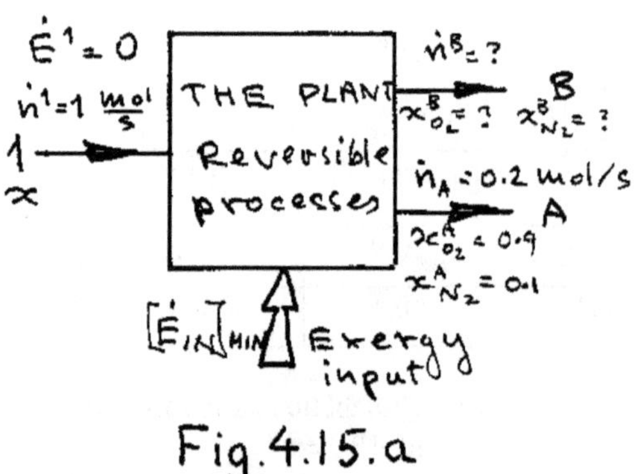

Fig.4.15.a

Assumptions: Atmospheric air consists of N_2 and O_2 only.

Analysis:

(i) Balance of substances

$$\dot{n}_{\mathrm{o2}}^{\mathrm{B}} = \dot{n}_{\mathrm{o2}}^{1} - \dot{n}_{\mathrm{o2}}^{\mathrm{A}} = 0.21\frac{\mathrm{mol}}{\mathrm{s}} - 0.18\frac{\mathrm{mol}}{\mathrm{s}} = 0.03\frac{\mathrm{mol}}{\mathrm{s}}$$

$$\dot{n}^{B}_{N2} = \dot{n}^{1}_{N2} - \dot{n}^{A}_{N2} = 0.79\frac{mol}{s} - 0.02\frac{mol}{s} = 0.77\frac{mol}{s}$$

Molar flow rate of stream B

$$\dot{n}^{B} = \dot{n}^{1} - \dot{n}^{A} = 1\frac{mol}{s} - 0.2\frac{mol}{s} = 0.8\frac{mol}{s}$$

Mole fractions

$$x^{B}_{o2} = \frac{\dot{n}^{B}_{o2}}{\dot{n}^{B}} = \frac{0.03}{0.8} = \underline{0.0375}$$

$$x^{B}_{N2} = \frac{\dot{n}^{B}_{N2}}{\dot{n}^{B}} = \frac{0.77}{0.8} = \underline{0.9625}$$

(ii) Since $P_A = P_B = P_o$ and $T_A = T_B = T_o$ $\dot{E}_A = \dot{E}^o_A$ and $\dot{E}_B = \dot{E}^o_B$

Chemical exergy rate of an ideal mixture

$$\dot{E}^o_M = \sum \dot{n}_i \tilde{\varepsilon}^o_i + \tilde{R}T^o \sum_i \dot{n}_i \ln x_i \qquad \text{(a)}$$

But for a reference substance

$$\tilde{\varepsilon}^o_i = \tilde{R}T^\circ \ln \frac{1}{x_{ioo}} \qquad \text{(b)}$$

Substituting (a) in (b)

$$\dot{E}^o = \tilde{R}T^\circ \sum_i n_i \ln \frac{x_i}{x_{ioo}} \qquad \text{(c)}$$

From an exergy balance for the plant with $\dot{E}^1 = 0$

$$\lfloor \Delta\dot{E}_{IN} \rfloor_{MIN} = \dot{E}_A + \dot{E}_B \qquad \text{(d)}$$

Substituting numerical data into (c)

$$\dot{E}_A = 8.3143 \times 293 \left[0.18 \ln \frac{0.9}{0.21} + 0.02 \ln \frac{0.1}{0.79} \right]$$

$$= 537.4 \text{ W}$$

$$\dot{E}_B = 8.3143 \times 293 \left[0.03 \ln \frac{0.0375}{0.21} + 0.77 \ln \frac{0.9625}{0.79} \right]$$

$$= 244.566 \text{ W}$$

Hence

$$(\Delta\dot{E}_{IN})_{MIN} = 537.4 + 244.566$$

$$= \underline{781.96 \text{ W}}$$

Determination of the minimum exergy of separation from the $\tilde{\varepsilon} - x$ chart

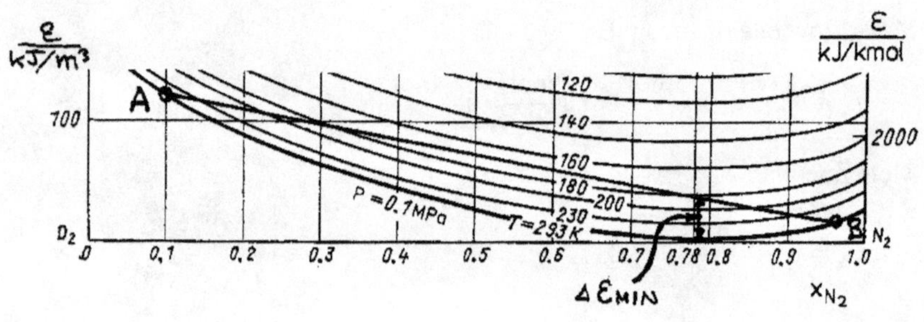

Fig. 4.15.b

Using the right-hand scale $\Delta\varepsilon_{\mathrm{MIN}} = 790 \dfrac{\mathrm{kJ}}{\mathrm{kmol}}$

Exergy input rate $\left(\Delta\dot{E}_{\mathrm{IN}}\right)_{\mathrm{MIN}} = \dot{n}\left(\Delta\varepsilon_{\mathrm{MIN}}\right)$

$$= 1\frac{\mathrm{mol}}{\mathrm{s}} 790 \frac{\mathrm{kJ}}{\mathrm{kmol}} = \underline{790\ \mathrm{W}}$$

Problem 4.16

The process: An exothermic chemical reaction of CO and H_2O vapour.

To determine: (i) The temperature of the product stream, and

(ii) irreversibility of the process per kmol of CO.

Given data: $T_1 = T_2 = 150\ °\mathrm{C}$

$P_1 = P_2 = P_3 = 1\ \mathrm{atm}$

$T_o = 25\ °\mathrm{C}$

$(H_2O)_g$ 1 → $CO + H_2O \rightarrow CO_2 + H_2$ → 3 $H_2 + CO_2$; CO 2

Fig. 4.16

Assumptions: The chamber is adiabatic and the reaction is complete, the products are an ideal mixture, $\Delta KE = 0$ and $\Delta PE = 0$.

Analysis: (i) Energy balance for a chemical reaction

$$\dot{Q} - \dot{W}_{\mathrm{x}} = \left[\sum_k \dot{n}_{\mathrm{k}} \tilde{h}^{\mathrm{o}}_{\mathrm{d,k}} - \sum_j \dot{n}_{\mathrm{j}} \tilde{h}^{\mathrm{o}}_{\mathrm{d,j}}\right] + \left[\sum_k \dot{n}_{\mathrm{k}} \tilde{h}_{\mathrm{ph,k}} - \sum_j \dot{n}_{\mathrm{j}} \tilde{h}_{\mathrm{ph,j}}\right] \tag{a}$$

For the present case, with $\dot{Q} = 0$, $\dot{W}_{\mathrm{x}} = 0$, $\tilde{h}^{\mathrm{o}}_{\mathrm{d,H_2O}} = 0$, $\tilde{h}^{\mathrm{o}}_{\mathrm{d,CO_2}} = 0$, we can write it as

$$\left[\tilde{h}^{\mathrm{o}}_{\mathrm{d}} + \tilde{c}^{\mathrm{h}}_{P}(T_2 - T_o)\right]_{\mathrm{CO}} + \left[\tilde{c}^{\mathrm{h}}_{P}(T_1 - T_o)\right]_{\mathrm{H_2O}}$$

$$= \left[\tilde{h}^{\mathrm{o}}_{\mathrm{d H_2}} + \left((\tilde{c}^{\mathrm{h}}_{P\mathrm{H2}} + \tilde{c}^{\mathrm{h}}_{P\mathrm{CO_2}}\right)(T_3 - T_o)\right] \tag{b}$$

Solving for T_3

$$T_3 = T_o + \frac{\left[\left(\tilde{h}_d^o\right)_{CO} - \left(\tilde{h}_d^o\right)_{H_2}\right] + \left(\tilde{c}_{P,CO}^h + \tilde{c}_{P,H_2O}^h\right)\left(T_3 - T^o\right)}{\tilde{c}_{P,H_2}^h + \tilde{c}_{P,CO_2}^h} \qquad \text{(c)}$$

A trial value of T_3 = 700 °C is first used.

From Tables A3 and D1

	H_2O	CO	H_2	CO_2	Remarks
$\tilde{h}_d^o$ /[kJ/kmol]	0	283150	242000	0	
$\tilde{c}_p^h$ /[kJ/kmolK]	33.64	29.44	-	-	at 150 °C
$\tilde{c}_p^h$ /[kJ/kmolK]	-	-	29.87	48.08	at 700 °C
$\tilde{c}_p^h$ /[kJ/kmolK]	-	-	29.79	47.62	at 650 °C

Substituting in (c)

$$T_3 = \frac{283150 - 242000 + (33.64 + 29.44)(150 - 25)}{29.87 + 48.08} = 654\,°\text{C}$$

As this result is too far from the assumed value of 700 °C, another value is tried, T_3 = 650 °, and the result obtained is

$\underline{T_3 = 658\ °\text{C}}$

(ii) Exergy balance

$$I = \tilde{\varepsilon}_{H_2O} + \tilde{\varepsilon}_{CO} - 2\tilde{\varepsilon}_M \qquad \text{(d)}$$

From Tables A3 and D3

	H_2O	CO	H_2	CO_2	Remarks
$\tilde{\varepsilon}^o$ /[kJ/kmol]	11710	275430	238490	20140	
$\tilde{c}_p^\varepsilon$ /[kJ/kmolK]	5.52	4.87	-	-	at 150 °C
$\tilde{c}_p^\varepsilon$ /[kJ/kmolK]	-	-	13.92	22.92	at 658 °C

$$\tilde{\varepsilon}_{H_2O}^o = \tilde{\varepsilon}_{H_2O}^o + \tilde{c}_{PH_2O}^\varepsilon (T_1 - T_o) = 11710 + 5.52 \times 125 = 12400\ \text{kJ/kmol}$$

Similarly,

$$\tilde{\varepsilon}_{CO}^o = 275430 + 4.78 \times 125 = 276039\ \text{kJ/kmol}$$

For the mixture

$$\tilde{\varepsilon}_M = (\tilde{\varepsilon}_{ph})_M + \tilde{\varepsilon}_M^o$$

where

$$(\tilde{\varepsilon}_{\text{ph}})_{\text{M}} = (T - T_{\text{o}})\sum_i x_{\text{i}}\tilde{c}^{\varepsilon}_{p,\text{i}} = 11660 \text{ kJ/kmol}$$

$$\tilde{\varepsilon}^{\text{o}}_{\text{M}} = \sum_i x_{\text{i}}\tilde{\varepsilon}^{\text{o}}_{\text{i}} + \tilde{R}T_{\text{o}}\sum_i x_{\text{i}} \ln x_{\text{i}} = 127597 \text{ kJ/kmol}$$

$$\tilde{\varepsilon}_{\text{M}} = 11660 + 127597 = 139257 \text{ kJ/kmol}$$

$\therefore \quad I = 12400 + 276039 - 2\times 139257 = \underline{9925 \text{ kJ/kmol}}$ of CO

Problem 4.17

The process: Steady state combustion of CH_4 with 120% of stoichiometric air.

To determine: (i) Exit temperature of the products and

(ii) process irreversibility.

Given data: $P_1 = P_2 = P_3 = P\text{o} = 1$ atm

$T_1 = T_2 = T_{\text{o}} = 25$ °C

Fig.4.17

Assumptions: The reaction is complete, the chamber is adiabatic, $\Delta KE = 0$ and $\Delta PE = 0$.

Analysis:

(i) Combustion equation with 20% air excess

$$CH_4 + 2.4\ O_2 + 9.024\ N_2 \rightarrow CO_2 + 2H_2O + 9.024\ N_2 + 0.4\ O_2 \qquad \text{(a)}$$

Since $\tilde{h}^{\text{o}}_{\text{d}}$ is zero for O_2, N_2 CO_2 and H_2O the energy balance is

$$\left(\tilde{h}^{\text{o}}_{\text{d}}\right)_{CH_4} = (T_3 - T_{\text{o}})\left[\left(\tilde{c}^{\text{h}}_P\right)_{CO_2} + 2\left(\tilde{c}^{\text{h}}_P\right)_{H_2O} + 9.024\left(\tilde{c}^{\text{h}}_P\right)_{N_2} + 0.4\left(\tilde{c}^{\text{h}}_P\right)_{O_2}\right] \qquad \text{(b)}$$

From Table A2 $\quad \left(\tilde{h}^{\text{o}}_{\text{d}}\right)_{CH_4} = 802320 \text{ kJ/kmol}$

and from Table D1 for $T_3 = 1800$ °C (trial value)

	CO_2	H_2O	N_2	O_2
$\tilde{c}^{\text{h}}_p$ /[kJ/kmolK]	54.21	41.86	33.14	34.66

Solving (b) for T_3 and substituting data

$$T_3 = 25 + \frac{802320}{54.21 + 2\times 41.86 + 9.024\times 33.14 + 0.4\times 34.66} = \underline{1804\ ^\circ\text{C}}$$

which is close enough to the trial value.

Exergy balance, with $\tilde{\varepsilon}_{\text{AIR}} = 0$

$$I = \tilde{\varepsilon}^{\text{o}}_{CH_4} - E_{\text{M}} \qquad \text{(c)}$$

where

$$E_M = (E_{ph})_M + (E^o)_M \qquad (d)$$

$$E_M^o = \sum_i n_i \tilde{\varepsilon}_i^o + \tilde{R}T^o \sum_i n_i \ln x_i \qquad (e)$$

With P = const,

$$(E_{ph})_M = (T_3 - T_o)\sum_i n_i \tilde{c}_{Pi}^{\varepsilon} \qquad (f)$$

Data from Tables A2 and D3

	CO_2	H_2O	N_2	O_2	CH_4
ε^o /[kJ/kmolK] 1800 °C	20140	11710	690	3970	836510
$\tilde{c}_p^{\varepsilon}$ /[kJ/kmolK]	37.66	28.84	22.71	23.78	-
x_i	0.0805	0.1610	0.728	0.032	

Substituting in (e), (f) and (d)

$$E_M^o = 20140 + 2\times 11710 + 9.024\times 690 + 0.4\times 3970$$

$$+ 8.3144\times 298.15[\ln 0.0805 + 2\ln 0.161 + 9.024\ln 0.728 + 0.4\ln 0.032]$$

$$= 25499 \text{ kJ/kmol } CH_4$$

$$\left(E_{ph}\right)_M = (1804 - 25)[37.66 + 2\times 28.84 + 9.024\times 22.71 + 0.4\times 23.78]$$

$$= 551111 \text{ kJ/kmol } CH_4$$

$\therefore$ E_M = 551111 + 25499 = 576610 kJ/kmol CH_4

Finally, substituting in (c)

I = 836510 - 5766110 = <u>259900 kJ/kmol CH_4</u>

Comment: Combustion is an inherently irreversible process. The irreversibility destroys in this example 31% of the exergy of the fuel.

Problem 4.18

The process: Pre-heating applied to the combustion process described in Problem 4.17.

To determine: (i) Exit temperature of the products,

(ii) process irreversibility and compare with the process without pre-heating.

Given data:

$$P_1 = P_2 = P_3 = Po = 1 \text{ atm}$$

$$T_1 = 80\ °C \qquad T_2 = 200\ °C \qquad T_o = 25\ °C$$

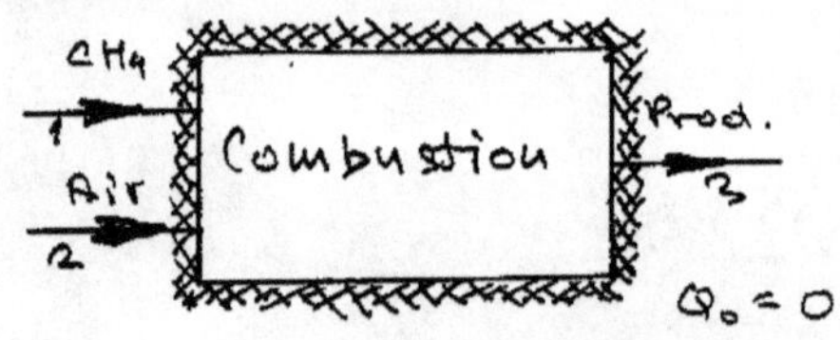

Fig.4.18

Assumptions: As in Problem 4.17.

Analysis:

Energy equation

$$\left(\tilde{h}_d^o\right)_{CH_4} + \left(\tilde{c}_P^h\right)_{CH_4} \times (T_1 - T_o) + 11.424\left(\tilde{c}_P^h\right)_{AIR} \times (T_2 - T_o)$$

$$= (T_3 - T_o)\left[\left(\tilde{c}_P^h\right)_{CO_2} + 2\left(\tilde{c}_P^h\right)_{H_2O} + 9.024\left(\tilde{c}_P^h\right)_{N_2} + 0.4\left(\tilde{c}_P^h\right)_{O_2}\right] \tag{a}$$

From Tables A2, D1 and D3

	CO_2	H_2O	N_2	O_2	80 °C CH_4	200 °C Air
$\tilde{c}_P^h$ /[kJ/kmolK] 1925 °C	54.65	42.37	33.34	34.80	37.22	29.47
x_i (products)	0.0805	0.1610	0.726	0.032		-
$\tilde{\varepsilon}^o$ /[kJ/kmol]	20140	11710	690	3970	836510	0
$\tilde{c}_P^h$ /[kJ/kmolK] 1925 °C	38.63	29.73	23.27	24.30	3.094	6.31

Solving a) for T_3 and substituting data

$$T_3 = 25 + \frac{802370 + 37.22(80-25) + 11.424 \times 29.47(200-25)}{54.65 + 2 \times 42.37 + 9.024 \times 33.24 + 0.4 \times 34.80}$$

$= \underline{1925\ °C}$

(N.B. this is the result of the final iteration.)

(ii) Process irreversibility from the exergy balance

$$I = \tilde{\varepsilon}_{CH_4} + n_{AIR}\tilde{\varepsilon}_{AIR} - E_M \tag{b}$$

$$\tilde{\varepsilon}_{\text{CH}_4} = \tilde{\varepsilon}^{\text{o}}_{\text{CH}_4} + \left(\tilde{c}_P^{\varepsilon}\right)_{\text{CH}_4}(T_1 - T_{\text{o}}) \tag{c}$$

$$\tilde{\varepsilon}_{\text{AIR}} = \left(\tilde{c}_P^{\varepsilon}\right)_{\text{AIR}} \times (T_2 - T_{\text{o}}) \tag{d}$$

E_{M} is calculated using expressions (d), (e) and (f) from the solution of Problem 4.17.

Substituting numerical data

$$\tilde{\varepsilon}_{\text{CH}_4} = 836680 \text{ kJ/kmol}$$

$$\tilde{\varepsilon}_{\text{AIR}} = 12615 \text{ kJ/kmol}$$

$$E^{\text{o}}_{\text{M}} = 25499 \text{ kJ/kmol}$$

$$(E^{\text{o}}_{\text{ph}})_{\text{M}} = 603817 \text{ kJ/kmol}$$

$\therefore \; E_{\text{M}} = 629316$ kJ/kmol

Hence, from (b)

$I = \underline{219975 \text{ kJ/kmol of CH}_4}$

Comment: As a result of preheating of the reactants, there is a 15.4% reduction in the process irreversibility. This is accompanied by a 91% increase in the exergy of the products of combustion which is due to their increased exit temperature. These improvements are achieved through the use of low grade thermal energy used in the preheating process.

Problem 4.19

The process: The combustion process described in Problem 4.18 is modified by limiting the products exit temperature through the use of a larger air-fuel ratio.

To determine: (i) The air excess necessary to limit the products exit temperature to ~1800 °C,
(ii) process irreversibility and
(iii) comment on the merits of the process.

Given data:

$P_1 = P_2 = P_3 = Po = 1$ atm

$T_1 = 80$ °C $\qquad T_2 = 200$ °C

$T_3 \cong 1800$ °C $\qquad T_{\text{o}} = 25$ °C

Assumptions: As in Problem 4.17.

Analysis:

Take $X = \dfrac{\text{actual air - fuel ratio}}{\text{stiochiometric air - fuel ratio}}$

Fig. 4.19

Reaction equation

$$CH_4 + 2XO_2 + 7.524X\,N_2 \rightarrow CO_2 + 2H_2O + 7.524X\,N_2 + 2(X-1)O_2 \qquad \text{(a)}$$

Energy equation

$$\left(\tilde{h}_d^o\right)_{CH_4} + \left(\tilde{c}_P^h\right)_{CH_4} \times (T_1 - T_o) + 9.524 \times \left(\tilde{c}_P^h\right)_{AIR} \times (T_1 - T_o)$$

$$= (T_3 - T_o)\left[\left(\tilde{c}_P^h\right)_{CO_2} + 2\left(\tilde{c}_P^h\right)_{H_2O} + 7.524X\left(\tilde{c}_P^h\right)_{N_2} + 2(X-1)\left(\tilde{c}_P^h\right)_{O_2}\right] \qquad \text{(b)}$$

Substituting numerical data in (b) and solving for X we get

$X = 1.32$

or air excess = $(X - 1)$ = 0.32

(ii)

	CO_2	H_2O	N_2	O_2	80 °C CH_4	200 °C Air
n_i /[kJ/kmol]	1	2	9.932	0.64	1	12.57
x_i	0.0737	0.1474	0.732	0.0472		
$\tilde{\varepsilon}_i^o$ /[kJ/kmol] at 1800 °C	20140	11710	690	3970	836510	0
$\tilde{c}_P^\varepsilon$ /[kJ/kmolK]	37.68	28.86	22.72	23.79	3.094	6.31

Exergy balance

$$I = \tilde{\varepsilon}_{CH_4} + n_{AIR}\tilde{\varepsilon}_{AIR} - E_M \qquad \text{(c)}$$

From expressions used in Problems 4.17 and 4.18, with the above data we obtain

$\tilde{\varepsilon}_{CH_4} = 836680$ kJ/kmol

$\tilde{\varepsilon}_{AIR} = 1104.2$ kJ/kmol

$E_M^o = 30594$ kJ/kmol

$(E_{ph})_M = 597571$ kJ/kmol

$\therefore\ E_M = 629316$ kJ/kmol

Hence, from (c)

I = 222395 kJ/kmol CH_4

Comment: This process, in comparison with combustion without preheating, gives a reduction in process irreversibility of 14.4% and an increase in the exergy of the products of 8.4%. Preheating of the reactants combined with an increase in the air excess give both benefits in improved performance and the advantage of a lower maximum temperature to meet metallurgical requirements and a reduction in NO_x.

Problem 4.20

The process: Adiabatic combustion of CO with different air excesses.

To determine: Adiabatic flame temperature (products exit temperature) and process irreversibility for the cases when air is supplied with (i) 30%, (ii) 100% excess over stoichiometric requirement.

Fig.4.20

Given data:

$$P_1 = P_2 = P_3 = Po = 1 \text{ atm}$$

$$T_1 = T_2 = T_o = 25\ °\text{C}$$

Analysis:

(i) 30% air excess

Combustion equation

$$\text{CO} + 1.3[0.5\,\text{O}_2 + 1.88\,\text{N}_2] \rightarrow \text{CO}_2 + 0.15\,\text{O}_2 + 2.444\,\text{N}_2 \tag{a}$$

From the energy equation

$$T_3 = T_o + \frac{\left(\tilde{h}_d^o\right)_{CO}}{\left(\tilde{c}_P^h\right)_{CO_2} + 0.15\left(\tilde{c}_P^h\right)_{O_2} + 2.444\left(\tilde{c}_P^h\right)_{N_2}} \tag{b}$$

Values of $\left(\tilde{c}_P^h\right)$ for $T_3 = 2025$ °C (trial value)

	Co_2	O_2	N_2
n_i/[kmol]	1	0.15	2.444
$\tilde{c}_P^h$/[kJ/kmolK]	54.97	34.91	33.50

Substituting in (b) together with $\left(\tilde{h}_d^o\right)_{CO} = 283150 \dfrac{\text{kJ}}{\text{kmol}}$ we get

$T_3 = 2018$ °C

which is close enough to the trial value.

Irreversibility from the exergy balance

$$I = \tilde{\varepsilon}_{CO}^o + E_{AIR} = E_M \tag{c}$$

where, with $E_{AIR} = 0$

$$E_M = \left(E_{ph}\right)_M + \left(E^o\right)_M \tag{d}$$

$$\left(E^o\right)_M = \sum_i n_i \tilde{\varepsilon}^o + \tilde{R}T^o \sum_i n_i \ln x_i \tag{e}$$

$$\left(E_{ph}\right)_M = (T_3 - T_o)\sum_i n_i \left(\tilde{c}_P^\varepsilon\right)_i \tag{f}$$

Data from tables A2 and D3

	CO_2	O_2	N_2
n_i/kmol	1	0.15	2.444
x_i	0.278	0.0417	0.680
$\tilde{\varepsilon}^{o}$/[kJ/kmol]	20140	3970	720
$\tilde{c}_P^{\varepsilon}$/[kJ/kmol]	39.35	24.68	23.69

$$\tilde{\varepsilon}_{CO}^{o} = 275430 \frac{kJ}{kmol}$$

Substituting numerical data in (e) and (f) we get

$$\left(E^{o}\right)_{M} = 15804 \text{ kJ/kmol CO}$$

$$\left(E_{ph}\right)_{M} = 201194 \text{ kJ/kmol CO}$$

Hence, from (d)

$$E_{M} = 216998 \text{ kJ/kmol CO}$$

and from (c)

$$I = 275430 - 216998$$

$$= \underline{58432 \text{ kJ/kmol of CO}}$$

(ii) 100% air excess

The combustion equations becomes

$$CO + 2[0.5\,O_2 + 1.88\,N_2] \rightarrow CO_2 + 0.5\,O_2 + 3.76\,N_2$$

Using $\tilde{c}_P^{h}$ data for a trial value of T_3 = 1800 °C in (b)

$$T_3 = 25 + \frac{283150}{54.21 + 0.5 \times 34.66 + 3.76 \times 33.14}$$

$$= 1469 \text{ °C}$$

As this is too far from the trial value of 1800 °C, a new trial value of T_3 = 1500 °C is used.

$$T_3 = 25 + \frac{283150}{53.03 + 0.5 \times 34.28 + 3.76 \times 32.6}$$

$$= 1494 \text{ °C}$$

As this is close to the new trial value, the iteration is stopped.

Using expressions (c) through to (f) and the following data

	CO_2	O_2	N_2
n_i/kmol	1	0.5	3.76
x_i	0.190	0.095	0.715
$\tilde{\varepsilon}^o$/[kJ/kmol]	20140	3970	720
$\tilde{c}_P^{\varepsilon}$/[kJ/kmol]	21.35	22.32	21.15

we get

$$E^o{}_M = 14671.3 \text{ kJ/kmol CO}$$

$$(E_{ph})_M = 164577.9 \text{ kJ/kmol CO}$$

$$\therefore \quad E_M = 179249.2 \text{ kJ/kmol CO}$$

Hence,

$$I = 275430 - 179249.2$$

$$= \underline{96180.8 \text{ kJ/kmol CO}}$$

Comment: The effect of increasing the air excess is to reduce the temperature of the products stream, which is due to its dilution by nitrogen and unused oxygen. This leads to a reduction in the exergy of the products stream and, as follows from expression (c), to a corresponding increase in process irreversibility.

Problem 4.21

The process: An idealized model of quicklime production.

To determine: Process irreversibility for heat input temperatures of (a) 1250 K and (b) 1450 K.

Given data $T_1 = T_o = 25$ °C $\qquad P_1 = P_2 = P_3 = Po = 1$ atm

$T_2 = T_3 = 200$ °C

For CaO at 200 °C

$\tilde{c}_P^h = 47.60$ kJ/kmolK $\qquad \tilde{c}_P^s = 45.59$ kJ/kmolK

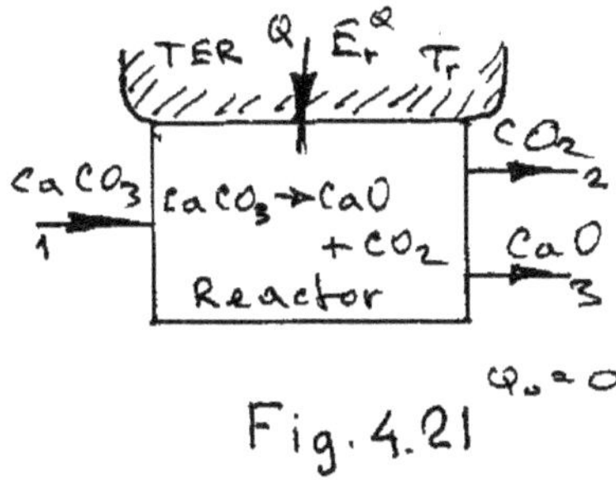

Fig. 4.21

Assumptions: Negligible heat losses to the environment, $\Delta KE = 0$ and $\Delta PE = 0$.

Analaysis:

Exergy balance for the reactor

$$E_r^Q = (E_{OUT} = E_{IN}) = I \qquad (a)$$

where (N.B. $(\tilde{\varepsilon}_{ph})_{CaCO_3} = 0$)

$$E_{OUT} - E_{IN} = (\tilde{\varepsilon}_{ph})_{CaO} + (\tilde{\varepsilon}_{ph})_{CO_2} + \tilde{\varepsilon}^0_{CaO} + \tilde{\varepsilon}^0_{CO_2} - \tilde{\varepsilon}^0_{CaCO_3} \qquad (b)$$

and

$$E_r^Q = Q_r \frac{T_r - T_o}{T_r} \qquad (c)$$

Energy balance, according to (A32), (N.B. $(\tilde{h}_{ph})_{CaCO_3} = 0$)

$$Q_r = (\tilde{h}_{ph})_{CaO} + (\tilde{h}_{ph})_{CO_2} + (\tilde{h}^o_d)_{CaO} + (\tilde{h}^o_d)_{CO_2} - (\tilde{h}^o_d)_{CaCO_3} \qquad (d)$$

From Table A3

	$CaCO_2$ 25 °C	CaO 200 °C	CO_2 200 °C
$(\tilde{h}^o_d)$/[kJ/kmol]	-170	177940	0
$\tilde{\varepsilon}^o$/[kJ/kmol]	6100	119620	20140

From CO_2 from Tables D1 and D2

$\tilde{c}_P^h = 40.52$ kJ/kmolK $\qquad \tilde{c}_P^\varepsilon = 9.09$ kJ/kmolK

Hence,

$$(\tilde{h}_{ph})_{CO_2} = \tilde{c}_P^h (T_2 - T_o) = 40.52 \times 175 = 7091 \text{ kJ/kmol}$$

$$(\tilde{\varepsilon}_{ph})_{CO_2} = \tilde{c}_P^\varepsilon (T_2 - T_o) = 9.09 \times 175 = 1590 \text{ kJ/kmol}$$

For CaO, using the given data

$$(\tilde{h}_{ph})_{CaO} = 47.60 \times 175 = 8330 \text{ kJ/kmol}$$

$$(\tilde{\varepsilon}_{ph})_{CaO} = \tilde{c}_P^h (T_3 - T_o) - T_o \tilde{c}_P^s \ln \frac{T_3}{T_o} = 47.60 \times 175 - 298.15 \times 45.59 \ln \frac{473.15}{298.15}$$

$$= 2053 \text{ kJ/kmol}$$

Substituting enthalpy values in (d) we get

$Q_r = 8330 + 7091 + 177940 + 0 + 170 = 193531$ kJ/kmol

Substituting exergy values in (b) we get

$$E_{\text{OUT}} - E_{\text{IN}} = 2053 + 1590 + 119620 + 20140 - 6100$$

$$= 142793 \text{ kJ/kmol CaCO}_3$$

Case (a)

$T_r = 1250$ K $\qquad \therefore E_r^Q = 193531 \dfrac{1250 - 298.15}{1250} = 147370 \text{ kJ/kmol CaCO}_3$

Substituting in (a)

$I = 147370 - 142793 = 4577$ kJ/kmol

$= \underline{4577 \text{ kJ/kmol CaCO}_3}$

Case (b)

$T_r = 1450$ K $\qquad \therefore E_r^Q = 193531 \dfrac{1450 - 298.15}{1450} = 153737 \text{ kJ/kmol CaCO}_3$

$\therefore$ $I = 153737 - 142793 = \underline{10944 \text{ kJ/kmol CaCO}_3}$

Comment: the higher is T_r above the minimum value here, when $I = 0$, $T_r = 1137$ K), the higher is the irreversibility (and, incidentally, also the speed of the reaction).

Problem 5.1

The plant: A coal-fired, steam electric power plant.

To determine: (i) Irreversibility rates of the main power plant components,

(ii) exergy flow rates,

(iii) rational efficiency of the plant, and

(iv) construct a Grassman diagram for the plant.

Given data:

$\dot{W}_{el} = 5000$ kW $\qquad P_1 = 15$ bar $\qquad T_1 = 300$ °C

$T_o = 10$ °C $\qquad P_{\text{COND}} = P_2 = P_3 = 0.05$ bar $\qquad T_{\text{COND}} = 306$ K

$\varphi = \dfrac{\varepsilon_F}{\text{NCV}} = 1.06 \qquad \eta_{\text{COMB}} = 0.80 \qquad \eta_s = 0.70 \qquad \eta_{el} = 0.90$

Assumptions: Feed pump power input, and pressure losses in the H_2O circuit are negligible. Assume $\dot{E}_3 = 0$, for the Grassmann diagram presentation.

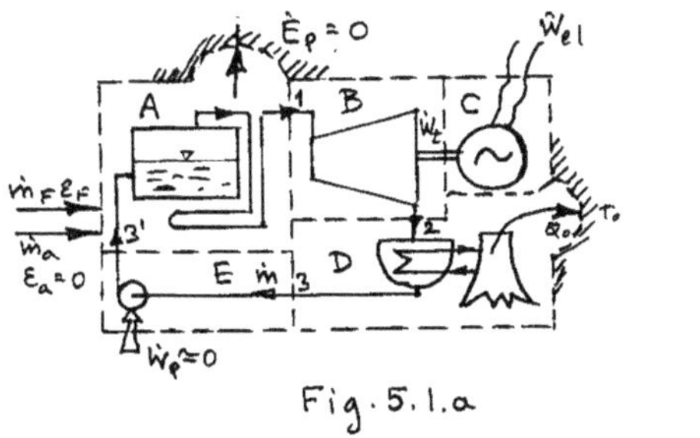

Fig. 5.1.a

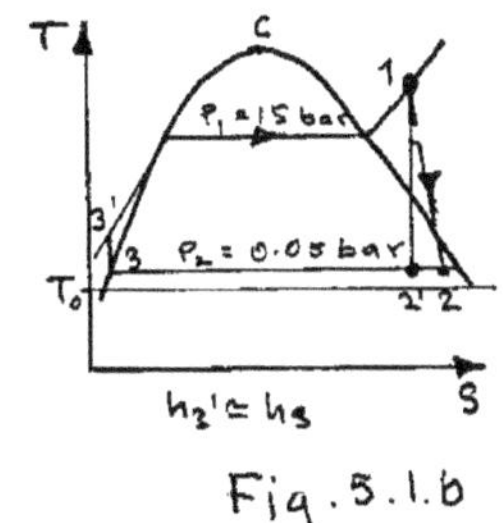

Fig. 5.1.b

Analysis:

Subregion A

Exergy balance

$$\dot{m}_f \varepsilon_F - \dot{m}(\varepsilon_1 - \varepsilon_3) = \dot{I}_A \qquad \text{(a)}$$

Combustion efficiency

$$\eta_{COMB} = \frac{\dot{m}(h_1 - h_3)}{\dot{m}_F(\text{NCV})} \qquad \text{(b)}$$

But $\varphi = \dfrac{\varepsilon_F}{(\text{NCV})}$ (c)

From (b) and (c)

$$\dot{m}_f \varepsilon_F = \varphi \dot{m}(h_1 - h_3)/\eta_{COMB} \qquad \text{(d)}$$

From (a) and (d)

$$i_A = \dot{I}_A / \dot{m} = \left(\frac{\varphi}{\eta_{COMB}} - 1\right)(h_1 - h_3) + T_o(s_1 - s_3) \qquad \text{(e)}$$

Subregion B

Exergy balance

$$\dot{m}(\varepsilon_1 - \varepsilon_2) - \dot{W}_t = \dot{I}_B \qquad \text{(f)}$$

but

$$\dot{W}_t = \dot{m}(h_1 - h_2) \qquad \text{(g)}$$

From (d) and (g)

$$\underline{i_B = \dot{I}_B / \dot{m} = T_o(s_2 - s_1)} \qquad \text{(h)}$$

Subregion C

Exergy balance

$$\dot{I}_C = \dot{W}_t - \dot{W}_{el} \qquad \text{(i)}$$

Using $\dot{W}_{el} = \eta_{el} W_t$ and (g), we get

$$\underline{i_C = \dot{I}_C / \dot{m} = (h_1 - h_2)(1 - \eta_{el})} \qquad \text{(j)}$$

Subregion D

From the exergy balance

$$\underline{i_D = \dot{I}_D / \dot{m} = (\varepsilon_2 - \varepsilon_3) = (h_2 - h_3) - T_o(s_2 - s_3)} \qquad \text{(k)}$$

Subregion E

Since $\dot{W}_p \cong 0$ $\quad \underline{\dot{I}_E = 0}$ (l)

Plant rational efficiency

By definition

$$\psi = \frac{\text{OUTPUT}}{\text{INPUT}} = \frac{\dot{W}_{el}}{\dot{m}_F \varepsilon_F} \tag{m}$$

Using (d), (g)

$$\psi = \frac{\eta_{el}\eta_{COMB}(h_1 - h_2)}{\varphi(h_1 - h_3)} \tag{n}$$

Alternatively,

$$\psi = 1 - \frac{\dot{m}\sum i_k}{\dot{m}_F \varepsilon_F}$$

Using (d)

$$\psi = 1 - \frac{n_{COMB}\sum i_k}{\varphi(h_1 - h_3)} \tag{o}$$

Steam properties (from tables)

$h_1 = 3039$ kJ/kg $\quad s_1 = 6.919$ kJ/kgK

$h_3 = 138$ kJ/kg $\quad s_3 = 0.476$ kJ/kgK

$h_{2'} = h_3 + T_{COND}(s_2 - s_3) = 2109.6$ kJ/kg

$h_2 = h_1 - \eta_s(h_1 - h_{2'}) = 2388$ kJ/kg

$x_2 = (h_2 - h_3)/h_{fg} = 0.9288$

$\therefore$ $s_2 = 7.83$ kJ/kgK

Specific irreversibilities & ψ calculation

Substituting in (e), (h) and (k)

$i_A = 2767.2$ kJ/kg

$i_B = 257.9$ kJ/kg

$i_C = 65.1$ kJ/kg

$i_D = 167.7$ kJ/kg

From either (n) or (o) we get the rational efficiency

$\underline{\psi = 0.152}$

Mass flow rate of steam

E1. power input $\quad \dot{W}_{el} = 5000$ kW

Turbine shaft power

$$\dot{W}_t = 5000/0.9 = 5555.6 \text{ kW}$$

Hence, from (g)

$$\dot{m} = \frac{5555.6}{3039 - 2388} = 8.53 \text{ kg/s}$$

Thus, irreversibility rates ($\dot{I}_k = \dot{m} i_k$)

	A	B	C	D	E	Total
$\dot{I}_k$ /kW	23604	2200	555	1430	0	27789.6

Taking feed water exergy as zero, i.e. $\dot{E}_3 = 0$
we get from the exergy balance for
Subregion D

$$\dot{E}_2 = \dot{I}_D = \underline{1430 \text{ kW}}$$

From the exergy balance for Sub region B

$$\dot{E}_1 = \dot{I}_B + \dot{W}_t + \dot{E}_2$$

$$= 2200 + 5555.6 + 1430 = \underline{9185.6 \text{ kW}}$$

From the exergy balance for Sub region A

$$\dot{E}_F = \dot{E}_1 + \dot{E}_3 + \dot{I}_A$$

$$= 9185.6 + 23604 = \underline{32789.6 \text{ kW}}$$

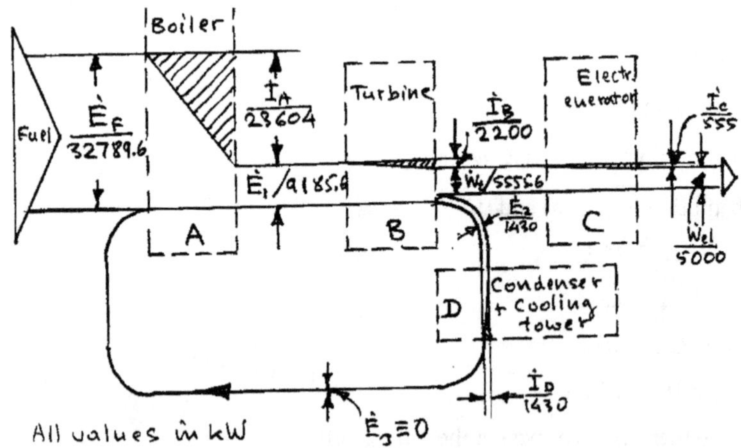

Fig.5.1.c

Comment: By far the largest irreversibility occurs in the boiler. This is an unavoidable form of irreversibility but its magnitude could be reduced by increasing pressure P_1 and temperature, T_1 of the steam generated and by improving the combustion efficiency. A further reduction could be achieved by using air preheating.

Problem 5.2

The plant: An ammonia refrigerator-heat pump plant for simultaneous freezing and heating.

To determine: the rational efficiency of the plant and construct a Grassmann diagram for it.

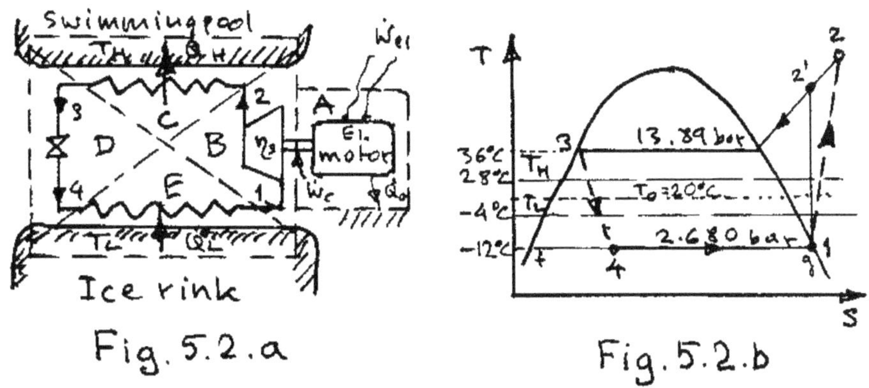

Fig. 5.2.a

Fig. 5.2.b

Given data:

$$\dot{Q}_{\mathrm{L}} = 50\,\mathrm{kW} \qquad \eta_{\mathrm{s}} = 0.82 \qquad \eta_{\mathrm{el}} = 0.9$$

Assumptions: Pressure losses in heat exchangers and stray heat transfer, $\Delta KE, \Delta PE$ are negligible.

Analysis:

Exergy balance for the whole system

$$\left(\dot{E}_{\mathrm{IN}} = \dot{E}_{\mathrm{OUT}} = 0\right) \qquad \dot{\mathrm{E}}^{\mathrm{Q}} = \dot{W}_{\mathrm{x}} + \dot{I} \tag{a}$$

where

$$\dot{E}^{\mathrm{Q}} = -\dot{E}_{\mathrm{H}}^{\mathrm{Q}} + \dot{E}_{\mathrm{L}}^{\mathrm{Q}} = -\dot{Q}_{\mathrm{H}} \frac{T_{\mathrm{H}} - T_{\mathrm{o}}}{T_{\mathrm{H}}} - Q_{\mathrm{L}} \frac{T_{\mathrm{o}} - T_{\mathrm{L}}}{T_{\mathrm{L}}} \tag{b}$$

and $\quad \dot{W}_{\mathrm{x}} = -\dot{W}_{\mathrm{el}}$ (c)

Substituting (b) and (c) in (a)

$$\underbrace{\dot{I} + \dot{W}_{\mathrm{el}}}_{\mathrm{INPUT}} = \underbrace{\dot{E}_{\mathrm{H}}^{\mathrm{Q}} + \dot{E}_{\mathrm{L}}^{\mathrm{Q}}}_{\mathrm{OUTPUT}} \tag{d}$$

Hence, the rational efficiency

$$\psi = \frac{\dot{E}_{\mathrm{H}}^{\mathrm{Q}} + \dot{E}_{\mathrm{L}}^{\mathrm{Q}}}{\dot{W}_{\mathrm{el}}} \tag{e}$$

or

$$\psi = 1 - \frac{\dot{I}}{\dot{W}_{\mathrm{el}}} \tag{f}$$

From the energy equation

$$\dot{W}_{\mathrm{C}} = \dot{m}(h_2 - h_1) \tag{g}$$

$$\dot{Q}_{\mathrm{H}} = \dot{m}(h_2 - h_3) \tag{h}$$

$$\dot{W}_{\mathrm{L}} = \dot{m}(h_1 - h_4) \tag{i}$$

Thermodynamic properties for NH_3

$h_1 = 1430.5$ kJ/kg $\qquad s_1 = 5.504$ kJ/kgK $= s_{2'}$

$h_3 = 352.3$ kJ/kg $\qquad s_3 = 1.298$ kJ/kgK

$= h_4 = h_f + x_4\,(h_g - h_f)$

$$\therefore x_4 = \frac{352.3 - 126.2}{1430.5 - 126.2} = 0.173$$

$$\therefore s_4 = 0.510 + 0.173(5.504 - 0.510)$$

$= 1.374$ kJ/kgK

By interpolation at 13.89 bar using $s_1 = s_{2'}$

$$h_{2'} = 1617.8 + (5.504 - 5.358)\,\frac{1745.7 - 1617.8}{5.692 - 5.358}$$

$= 1673.7$ kJ/kg

$$\therefore h_2 = h_1 + \frac{h_{2'} - h_1}{\eta_s} = 1727.1 \text{ kJ/kg}$$

and hence

$$s_2 = 5.538 + (5.692 - 5.358)\frac{1727.1 - 1617.8}{1745.7 - 1617.8}$$

$= 5.643$ kJ/kgK

From (i) $\qquad \dot{m} = \dfrac{50}{1430.5 - 352.3} = 0.0464\,\text{kg/s}$

Substituting in (g) and (h)

$$\dot{W}_{\mathrm{C}} = 13.76\,\text{kW} \quad \text{and} \quad \dot{\mathrm{Q}}_{\mathrm{H}} = 63.75\,\text{kW}$$

and hence

$$\dot{E}_{H}^{Q} = 63.75 \frac{301 - 293}{301} = 1.69 \text{ kW}$$

$$\dot{E}_{L}^{Q} = 50 \frac{293 - 269}{269} = 4.46 \text{ kW}$$

El. power $\qquad \dot{W}_{el} = \dfrac{\dot{W}_{C}}{\eta_{el}} = \dfrac{13.76}{0.9} = 15.29 \text{ kW}$

Rational efficiency

$$\psi = \frac{\text{OUTPUT}}{\text{INPUT}} = \frac{\dot{E}_{H}^{Q} + \dot{E}_{L}^{Q}}{\dot{W}_{el}} = \underline{0.402}$$

<u>Construction of the Grassmann diagram</u>

<u>Irreversibility rates</u>:

$$\dot{I}_{A} = \dot{W}_{el} - \dot{W}_{C} = 1.53 \text{ kW}$$

$$\dot{I}_{B} = \dot{m} T_{o} [s_{2} - s_{1}] = 1.89 \text{ kJ/kg}$$

$$\dot{I}_{C} = T_{o} \left[\dot{m}(s_{3} - s_{2}) + \frac{\dot{Q}_{H}}{T_{H}} \right] = 2.96 \text{ kW}$$

$$\dot{I}_{D} = \dot{m} T_{o} (s_{4} - s_{3}) = 1.03 \text{ kW}$$

$$\dot{I}_{E} = T_{o} [\dot{m}] (s_{1} - s_{4}) - \frac{\dot{Q}_{L}}{T_{L}} = 1.72 \text{ kW}$$

<u>Exergy flow rates</u>:

As assumed $\qquad \dot{E}_{1} = 0$

$$\dot{E}_{4} = \dot{E}_{L}^{Q} + \dot{I}_{E} = 4.46 + 1.72 = 6.18 \text{ kW}$$

$$\dot{E}_{3} = \dot{E}_{4} + \dot{I}_{D} = 6.18 + 1.03 = 7.21 \text{ kW}$$

$$\dot{E}_{2} = \dot{E}_{3} + \dot{E}_{H}^{Q} + \dot{I}_{C} = 7.21 + 1.69 + 2.96 = 11.86 \text{ kW}$$

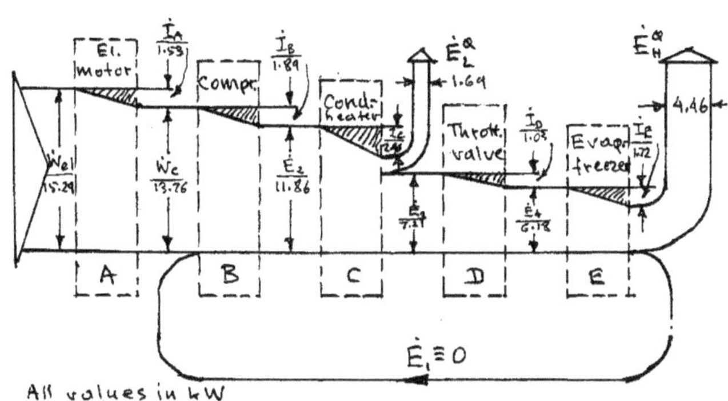

Fig.5.2.c

Comment: The largest irreversibility is in the condenser owing to the largest temperature differences in the de-superheating part of the heat exchange process.

Problem 5.3

The plant: Ammonia refrigerator with liquid sub-cooling.

To determine: Rational efficiency of the plant and construct a Grassmann diagram for it. Compare the rational efficiency with that of a plant without liquid sub-cooling.

Assumptions: Negligible pressure losses and stray heat transfers. $\Delta KE = 0, \Delta PE = 0$.

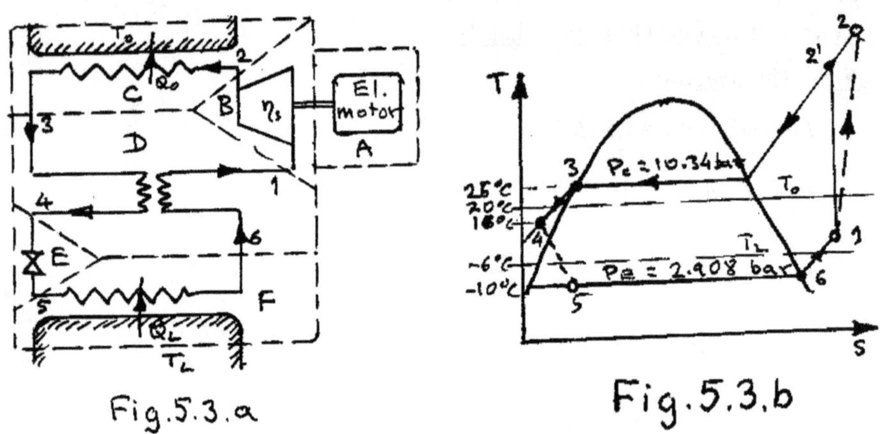

Fig.5.3.a Fig.5.3.b

Given data:

$\eta_s = 0.8 \qquad \eta_{el} = 0.9$

$\dot{Q}_L = 30\,\text{kW} \qquad T_o = 20\ °\text{C}$

$T_L = \text{-}6\ °\text{C} \qquad T_4 = 16\ °\text{C}$

For process $\quad 3 \rightarrow 4 \quad c_{Pe} = 4.75 \dfrac{\text{kJ}}{\text{kgK}}$

$\qquad\qquad\quad 6 \rightarrow 1 \quad c_{Pv} = 2.35 \dfrac{\text{kJ}}{\text{kgK}}$

Analysis:

Calculation of T_1: Energy balance for S-R D

$$c_{Pv}(T_1 - T_6) = c_{Pe}(T_3 - T_4) \tag{a}$$

$$\therefore \quad T_1 = -10 + \frac{4.75}{2.35}(26 - 16) = 10.2\ °\text{C} = 283.35\ \text{K}$$

For P = const

$$s_1 - s_6 + c_{Pv} \ln \frac{T_1}{T_6} = 5.649 \text{ kJ/kgK}$$

$$h_1 - h_6 + c_{Pv}(T_1 - T_6) = 1480.5 \text{ kJ/kgK}$$

Mass flow rate:

$$Q_L = \dot{m}(h_6 - h_5) \qquad h_5 = h_4 = 256.0 \text{ kJ/kg}$$

$$h_6 = 1433.0 \text{ kJ/kg}$$

$$\therefore \ \dot{m} = \frac{30}{1433 - 256} = \underline{0.0255 \text{ kg/s}}$$

Expression for relative exergy

$$\dot{E} = \dot{m}(h - T_o s) - \beta_r \qquad \text{(b)}$$

It is assumed here $\dot{E}_1 = 0$, hence

$$\beta_r = (h_1 - T_o s_1) = \underline{-175.5 \text{ kJ/kg}}$$

Subregions A and B

$$\dot{W}_{el} = \dot{W}_C / \eta_{el} \qquad \text{(c)}$$

$$\dot{W}_C = \dot{m}(h_2 - h_1) \qquad \text{(d)}$$

Since $s_1 = s_{2'}$

By interpolation at 10.34 bar $h_{2'} = 1676.8$ kJ/kg

For the compressor

$$h_2 = h_1 + \frac{h_{2'} - h_1}{\eta_s} = 1725.9 \text{ kJ/kg}$$

Using (d) $\dot{W}_C = 6.257$ kW

From (c) $\dot{W}_{el} = 6.95$ kW

By interpolation at 10.34 bar

$$s_2 = 5.780 \text{ kJ/kgK}$$

Exergy balance for Subregion A

$$\dot{I}_A = \dot{W}_{el} - \dot{W}_C = \underline{0.693 \text{ kW}}$$

Using (b) $\dot{E}_2 = 0.0255[1725 - 293.15 \times 5.780 + 175.5]$

$$= \underline{5.278 \text{ kW}}$$

Exergy balance for Subregion B

$$\dot{I}_B = \dot{E}_1 - \dot{E}_2 + \dot{W}_C$$

$$= 0 - 5.278 + 6.257 = \underline{0.979 \text{ kW}}$$

Subregion C

$\dot{I}_C = \dot{E}_2 - \dot{E}_3 + \dot{E}_o^Q$ $\quad$ h_3 = 303.7 kJ/kg

with $\dot{E}_D^Q = 0$ $\quad$ s_3 = 1.140 kJ/kgK

Using (b)

$\dot{E}_3 = 3.7\,\text{kW}$

$\dot{I}_C = 1.58\,\text{kW}$

Subregion D

$\dot{I}_D = \dot{E}_3 + \dot{E}_6 - (\dot{E}_4 + \dot{E}_1)$

Using (b) $\quad$ h_4 = 256.0 kJ/kg

$\underline{\dot{E}_4 = 3.685\,\text{kW}}$ $\quad$ s_4 = 0.979 kJ/kgK

$\underline{\dot{E}_6 = 0.0893\,\text{kW}}$ $\quad$ h_6 = 1433 kJ/kg

s_6 = 5.475 kJ/kgK

$\therefore \quad \dot{I}_D = 3.70 + 0.0893 - 3.685 = \underline{0.102\text{ kW}}$

Subregion E

$\dot{I}_E = \dot{E}_4 - \dot{E}_5$

Using (b) $\quad$ $h_5 = h_4 = h_f + x_5\,(h_g - h_f)$

$\dot{E}_5 = 3.51\,\text{kW}$

$$\therefore x_5 = \frac{256 - 135.4}{1433 - 135.4} = 0.093$$

$$\therefore s_5 = 0.544 + 0.093 \times 4.931$$

= 1.0026 kJ/kgK

$\dot{I}_E = 3.685 - 3.51 = \underline{0.176\text{ kW}}$

Subregion F

$\dot{I}_F = \dot{E}_5 - \dot{E}_6 + \dot{E}_L^Q$

where $\dot{E}_L^Q = \dot{Q}_L \dfrac{T_L - T_o}{T_L} = -30\dfrac{293.15 - 267.15}{167.15}$

= -2.9197 kW

Hence

$\dot{I}_F = \underline{0.501\text{ kW}}$

Rational efficiency

$$\psi = \frac{\dot{E}^{Q}}{\dot{W}_{el}} = 0.42$$

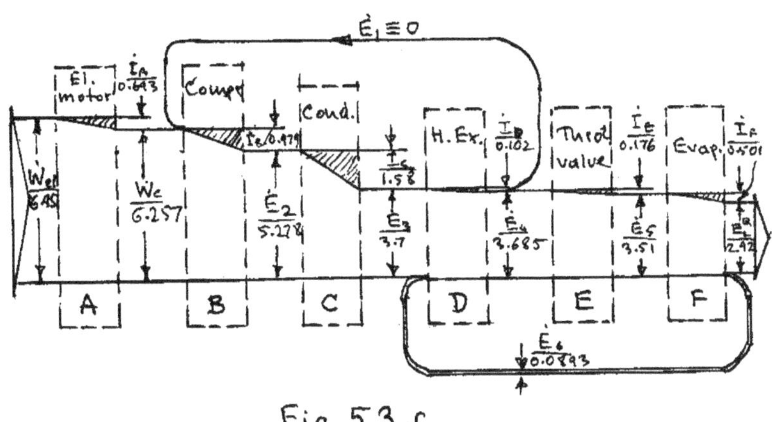

Fig. 5.3.c

Plant version without sub-cooling

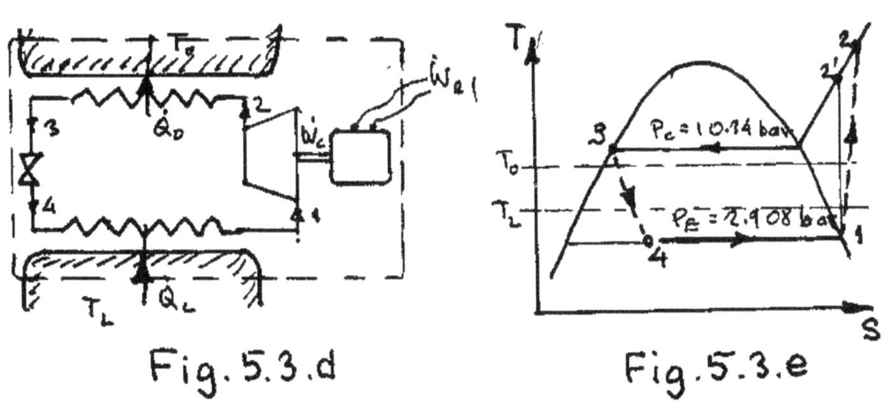

Fig. 5.3.d

Fig. 5.3.e

Properties of NH_3

$T_o = 20\ °C$ $T_L = -6\ °C$

$h_1 = 1433$ kJ/kg $s_1 = 5.475$ kJ/kgK

$h_3 = 303.7$ kJ/kg $s_3 = 1.140$ kJ/kgK

By interpolation at 10.34 bar ($s_1 = s_{2'}$)

$$h_{2'} = 1611.7\,\text{kJ/kg} \;\therefore h_2 = h_1 + \frac{h_2 - h_{2'}}{\eta_s} = 1656\ \text{kJ/kg}$$

$s_2 = 5.593$ kJ/kgK

$h_3 = h_4,\ s_4 = 1.184$ kJ/kgK

Mass flow rate:

$$\dot{m} = \frac{30}{1433 - 303.7} = 0.02656\,\text{kg/s}$$

Compressor power input

$$\dot{W}_C = 0.2656(1656 - 1433) = 5.924\,\text{kW}$$

$$\therefore\ \dot{W}_{el} = \dot{W}_C / \eta_{el} = 6.58\,\text{kW}$$

Coefficient of performance

$$(CP)_{REF} = \frac{\dot{Q}_L}{W_{el}} = \frac{30}{6.58} = 4.56$$

Hence

$$\psi = \frac{(CP)_{REF}}{T_L/(T_o - T_L)} = \underline{0.444}$$

Comment: Sub-cooling produces only a small increase in refrigeration duty (about 4%) but leads to a large increase in the compressor power input and consequently to a lower rational efficiency.

Problem 5.4

The plant: The Heylandt nitrogen liquefaction plant.

To determine: The rational efficiency of the plant and construct a Grassmann diagram and a $\tau - \dot{H}$ diagram for the heat exchange processes.

Given data: $\dot{m} = 0.1\,\text{kg/s}$

$P_o = 1.013$ bar $\qquad T_o = 300$ K

$P_1 = 202.6$ bar $\qquad T_1 = T_o$

$F = 0.55$ $\qquad T_8 = 297$ K

Compressor: $\eta_{iso} = 0.85;\quad \eta_m = 0.88$

Expander: $\eta_s = 0.75;\quad \eta_m = 0.9$

El. Motor overall efficiency $\eta_{el} = 0.95$

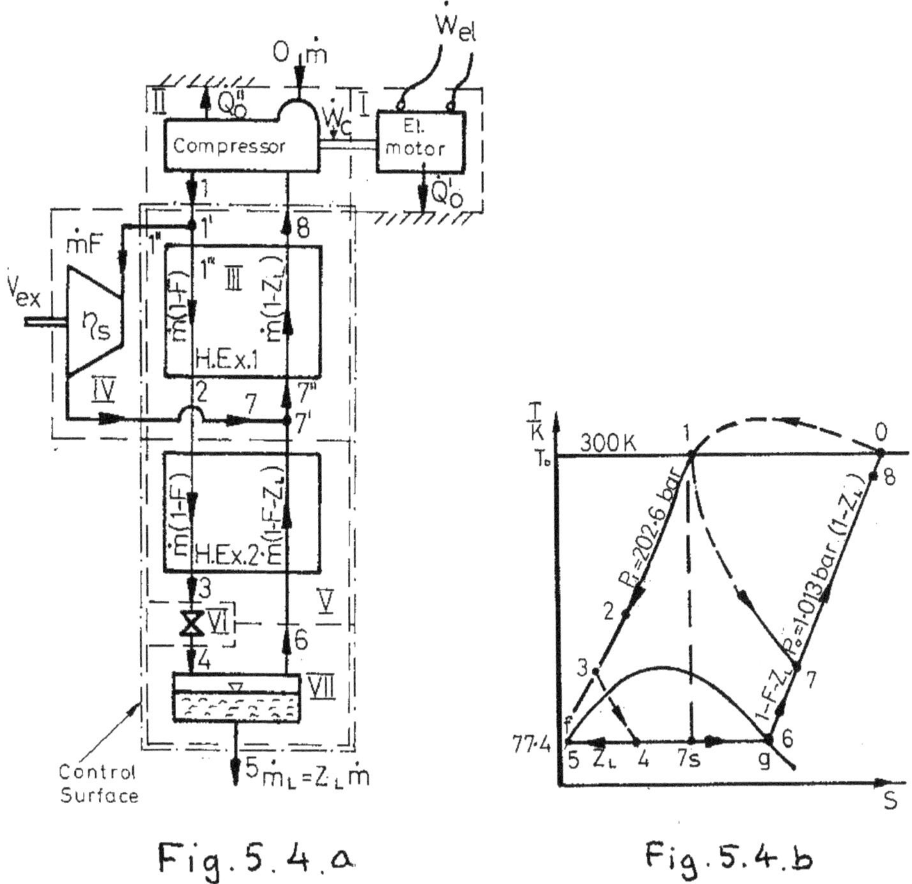

Fig.5.4.a

Fig.5.4.b

Assumptions: Pressure losses, stray heat transfer, ΔKE and ΔPE are negligible.

Analysis:

Energy balance for the control region defined by the double dotted line

$$\dot{m}h_1 + \dot{m}Fh_7 = \dot{m}Fh_1 + Z_L\dot{m}h_5 + \dot{m}(1 - Z_L)h_8$$

Hence,

$$Z_L = \frac{h_8 - h_1}{h_8 - h_5} + F\frac{h_1 - h_7}{h_8 - h_5} \qquad \text{(a)}$$

For the expander with isentropic efficiency η_s

$$h_3 = h_1 - \eta_s(h - h_{7s}) \qquad \text{(b)}$$

From the energy balance for S-R VII

$$h_4 = \frac{2_L h_5 + (1 - F - Z_L)h_6}{1 - F} \qquad \text{(c)}$$

Now, since the separation of the two phases is reversible, $\dot{I} = 0$. Hence, from the Gouy-Stodola relation, with $Q = 0$, $\dot{S}_{\text{IN}} - \dot{S}_{\text{OUT}}$

$$\therefore \quad s_4 = \frac{Z_L s_5 + (1 - F - Z_L) s_6}{1 - F} \quad \text{(d)}$$

For the throttling process

$$h_4 = h_3 \quad \text{(e)}$$

From the energy balance for S-R V

$$h_2 = h_3 + \frac{(1 - F - Z_L)(h_7 - h_6)}{1 - F} \quad \text{(f)}$$

For the compressor

$$\dot{W}_{\text{iso}} = \dot{m} R T_o \ln(P_1 / P_o) \quad \text{(g)}$$

$$W_{\text{ind}} = W_{\text{iso}} / \eta_{\text{iso}} \quad \text{and} \quad W_C = W_{\text{ind}} / \eta_m \quad \text{(h)}$$

El. Motor $\quad \dot{W}_{\text{el}} = \dot{W}_C / \eta_{\text{el}}$ (i)

Expander $\quad \dot{W}_{\text{ex}} = \dot{m} F (h_1 - h_7) \eta_m$ (j)

Thermodynamic properties of nitrogen

	$P = 1.013$ bar		P=206.6 bar	
$\frac{T}{K}$	$\frac{h}{\text{kJ/kg}}$	$\frac{s}{\text{kJ/kgK}}$	$\frac{h}{\text{kJ/kg}}$	$\frac{s}{\text{kJ/kgK}}$
77.4 sat.liq.	29.4	0.418		
77.4 sat. vap.	118.7	2.994		
100	253.0	3.270	87.2	0.784
110	263.6	3.371	106.3	0.966
120	274.2	3.463	124.5	1.124
260	420.5	4.272	376.9	2.548
280	441.3	4.349	404.8	2.652
300	462.1	4.421	431.5	2744

Thermodynamic properties

$h_1 = 431.5$ kJ/kg $\qquad s_1 = 2.744$ kJ/kgK

$h_o = 462.1$ kJ/kg $\qquad s_o = 4.421$ kJ/kgK

Since $s_1 = s_{7s} = s_f = + x_{7s} (s_g - s_f)$

$$x_{7s} = \frac{2.744 - 0.418}{2.994 - 0.418} = 0.903$$

h_{7s} = 29.4 + 0.903 (228.7 - 29.4) = 264.9 kJ/kg

From (b)

h_7 = 431.5 - 0.75 (431.5 - 209.4) = 264.9 kJ/kg

By interpolating in the table at P_o = 206.6 bar

T_7 = 111.2 K s_7 = 3.382 kJ/kgK

For point (8), T_8 = 297 K as assumed

h_8 = 459.0 kJ/kg s_8 = 4.410 kJ/kgK

From (a) Z_L = 0.2773

From (c) h_4 = 105.89 kJ/kg = h_3

From (d) s_4 = 1.407 kJ/kgK

By interpolating at P_1 = 202.6 bar

s_3 = 0.962 kJ/kgK

From (f) h_2 = 119.8 kJ/kg

Interpolating in the table at P_1 = 202.6 bar

s_2 = 1.083 kJ/kgK

T_2 = 117.4 K

T_3 = 109.8 K

As given

T_6 = 77.4 K

<u>Exergy flow rates</u>:

Using physical exergy values

$$\dot{E} = \dot{m}[(h - T_o s) - \beta_o]$$

where

$$\beta_o = h_o - T_o s_o = 462.1 - 300 \times 4.421$$

$$= -864.2 \text{ kJ/kg}$$

Thus

$$\dot{E}_1 = \dot{m}[(h_1 - T_o s_1) - \beta_o] = 47.25 \text{ kW}$$

$$\dot{E}_{1'} = (1 - F)\dot{E}_1 = 25.99 \text{ kW}$$

$$\dot{E}_2 = \dot{m}(1 - F)[(h_2 - T_o s_2) - \beta_o] = 29.66 \text{ kW}$$

$$\dot{E}_3 = \dot{m}(1 - F)[(h_3 - T_o s_3) - \beta_o] = 30.67 \text{ kW}$$

$$\dot{E}_4 = \dot{m}(1 - F)[(h_4 - T_o s_4) - \beta_o] = 24.66\,\text{kW}$$

$$\dot{E}_5 = \dot{m}Z_L[(h_5 - T_o s_5) - \beta_o] = 21.30\,\text{kW}$$

$$\dot{E}_6 = \dot{m}(1 - F - Z_L)[(h_6 - T_o s_6) - \beta_o] = 3.36\,\text{kW}$$

Specific exergy at point (7)

$$\varepsilon_7 = h_7 - T_o s_7 - \beta_o = 114.5\,\text{kJ/kg}$$

Since mixing of the streams at point (7) is assumed to be reversible

$$\dot{E}_7 = F \text{ in } \varepsilon_7 = 6.30\,\text{kW}$$

$$\dot{E}_{7'} = \dot{m}(1 - F - Z_L)\varepsilon_7 = 1.98\,\text{kW}$$

$$\dot{E}_{7''} = \dot{m}(1 - Z_7)\varepsilon_7 = 8.28\,\text{kW}$$

$$\dot{E}_8 = \dot{m}(1 - Z_L)[(h_8 - T_o s_8) - \beta_o] = 0.014\,\text{kW}$$

Power inputs and outputs

Using $R = 0.297$ kJ/kgK, we have from (g)

$$\dot{W}_{iso} = \underline{47.2\text{ kW}}$$

from (h) $\dot{W}_C = W_{iso}/(\eta_{iso} \times \eta_m) = \underline{63.10\text{ kW}}$

from (i) $\dot{W}_{el} = \underline{66.42\text{ kW}}$

from (j) $\dot{W}_{ex} = \underline{8.25\text{ kW}}$

Irreversibility rates - from the exergy balance

Subregion I, E1. Motor

$$\dot{I}_I = \dot{W}_{el} - \dot{W}_C = 66.42 - 63.10 = \underline{3.32\text{ kW}}$$

Subregion II, Compressor

$$\dot{I}_{II} = \dot{W}_C - \dot{E}_1 + \dot{E}_8 = 63.10 - 47.25 + 0.014 = \underline{15.86\text{ kW}}$$

Subregion III, Heat Exchanger 1

$$\dot{I}_{III} = \dot{E}_{1'} + \dot{E}_{7''} - \dot{E}_2 - \dot{E}_8$$

$$= 25.99 + 8.275 - 29.66 - 0.014 = \underline{4.59\text{ kW}}$$

Subregion IV, Expander

$$\dot{I}_{IV} = \dot{E}_{1''} - \dot{W}_{ex} - \dot{E}_7$$

$$= 25.99 - 8.25 - 6.30 = \underline{11.44\text{ kW}}$$

Subregion V, Heat Exchanger 2

$$\dot{I}_{\text{V}} = \dot{E}_2 + \dot{E}_6 - \dot{E}_3 - \dot{E}_{7'}$$

$$= 29.66 + 3.36 - 30.67 - 1.98 = \underline{0.37 \text{ kW}}$$

Subregion VI, Throttling valve

$$\dot{I}_{\text{VI}} = \dot{E}_3 - \dot{E}_4 = 30.67 - 24.66 = \underline{6.01 \text{ kW}}$$

τ Values

$$\tau_1 = \frac{T_1 - T_o}{T_1} = 0 \quad (\text{since } T_1 = T_o); \qquad \tau_2 = \frac{T_2 - T_o}{T_2} = -1.56$$

$$\tau_3 = \frac{T_3 - T_o}{T_3} = -1.73; \qquad \tau_6 = \frac{T_6 - T_o}{T_6} = -2.88$$

$$\tau_7 = \frac{T_7 - T_o}{T_7} = -1.70; \qquad \tau_8 = \frac{T_8 - T_o}{T_8} = -0.01$$

Heat transfer duties

Heat Exchanger 1 - Subregion III

$$\dot{Q}_{\text{III}} = \dot{m}(1 - F)(h_1 - h_2) = \underline{14.02 \text{ kW}}$$

Heat Exchanger 2 - Subregion V

$$\dot{Q}_{\text{V}} = \dot{m}(1 - F)(h_2 - h_3) = \underline{0.63 \text{ kW}}$$

Plant rational efficiency

$$\psi = \frac{\text{OUTPUT}}{\text{INPUT}}$$

Output = $\dot{E}_5 + \dot{W}_{\text{ex}} = 21.30 + 8.25 = 29.55$ kW

Input = $\dot{W}_{\text{el}} = 66.4$ kW

Hence,

$$\psi = \frac{29.55}{66.4} = \underline{0.445}$$

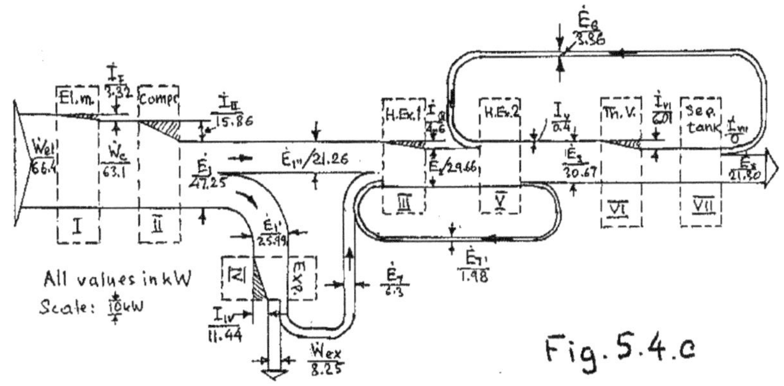

Fig. 5.4.c

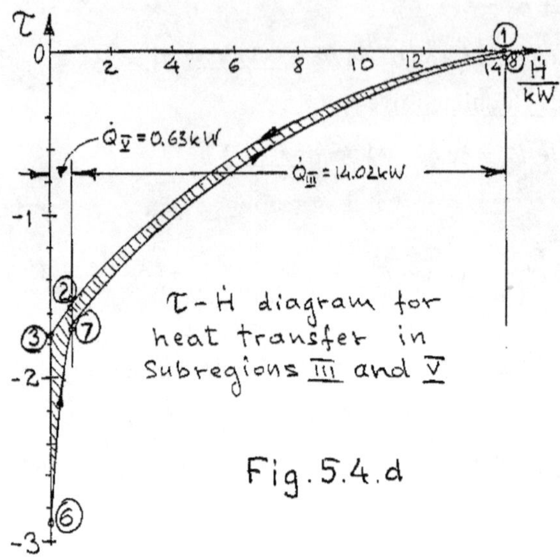

Fig. 5.4.d

Comment: As shown on the $\tau - \dot{H}$ diagram, the use of an expander in the Heylandt plant leads to a very low irreversibility rate in the heat exchangers, shown as the cross-hatched area on the diagram. By using a near-optimum value of F (0.55) the heat capacity rates of the two streams in Sub region III are very close in magnitude, which results in the two curves on the $\tau - \dot{H}$ diagram running nearly parallel. This permits the throttling valve inlet temperature, T_{IN}, to be reduced to a low value which gives a high proportion of the liquid (1 - x_4) at the end of the throttling process, as well as a low value of the process irreversibility, $\dot{I}_{VI}$. Although only (1-F) = 0.45 fraction of nitrogen goes through the throttling process, the liquid fraction, z_L, is higher than in the case of the other two liquefaction processes. A comparison of the different performance parameters is given in the table below. It should, however, be kept in mind that in the first two processes air is liquified whilst in the case of the Heylandt process, it is nitrogen.

Process	ψ	T_{IN}/K	Liquid Fraction (1-x_4)	Z_L
Simple Linde	0.0662	174	0.07	0.07
Linde + auxil. Refrig.	0.2198	155	0.24	0.24
Heylandt	0.445	109.8	0.62	0.28

Problem 5.5

The plant: An oil-fired steam boiler with fuel and air preheating.

To determine: Temperatures T_{F2}, T_{G4} and T_{G1}, irreversibilities per 1 kg of the fuel in the five sub-regions indicated, the rational efficiency and construct a Grassmann diagram.

Given data: $T_{A1} = T_{F1} = T_o = 25\ °C$ $\quad$ $T_{s1} = 25\ °C$

$T_{s2} = 500\ °C$ $\quad$ $P_s = 150$ bar = const

$P_A = P_F = P_G = P_o = 1$ atm $\quad$ Air excess = 30%

Fuel: $(NCV) = 42500$ kJ/kg, $c_{PF} = 202$ kJ/kgK

Composition: C -87% H - 13%

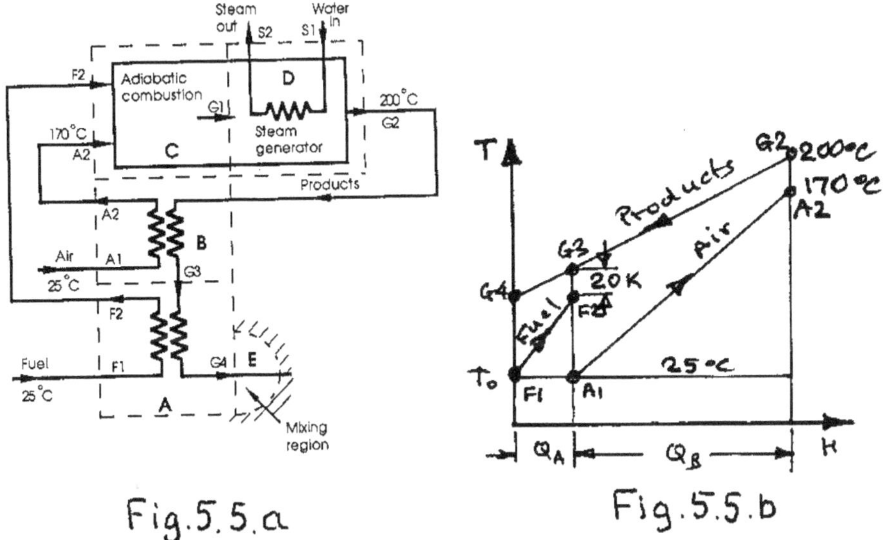

Fig.5.5.a $\qquad$ Fig.5.5.b

Assumptions: No stray heat transfer and pressure losses, $\Delta KE = 0, \Delta PE = 0$.

Analysis:

Combustion equations by mass:

Carbon: $\quad 0.87 \text{ kg C} + 2.32 \text{ kg } O_2 \rightarrow 3.19 \text{ kg } CO_2$

Hydrogen: $\quad 0.13 \text{ kg } H_2 + 1.04 \text{ kg } O_2 \rightarrow 1.17 \text{ kg } H_2O$

$$\text{Stoichiometric Air/Fuel} = \frac{2.32 + 1.04}{0.233} = 14.42$$

Actual $\quad$ Air/Fuel = $1.3 \times 14.42 = 18.75$

Products information

	N_2	O_2	CO_2	H_2O	Total
kg/kg of fuel	14.378	1.008	3.19	1.17	19.75
Mass fraction	0.728	0.051	0.1615	0.059	1
Molar mass $\frac{\tilde{m}}{\text{kg/kmol}}$	28	32	44	18	$\sum x_k \tilde{m}_k = 28.9$
n_k /[kmol/kg of fuel]	0.5135	0.0315	0.0725	0.065	0.6825
Mole fraction x_k	0.752	0.0462	0.106	0.095	1
$\tilde{c}_P$ /[kJ/kmolK] at 120 °C	29.35	28.20	39.70	33.50	$\sum x_k \tilde{c}_{Pk} = 30.76$

Specific heat capacities:

Gases: $c_{PG} = 30.76/38.9 = 1.06$ kJ/kgK

Fuel oil: $c_{PF} = 2.02$ kJ/kgK (as given)

Air: $c_{PA} = 1.005$ kJ/kgK

Energy balances for the air and fuel preheaters

$$\text{S-R B:} \quad \dot{m}_G c_{PG} = (T_{G2} - T_{G3}) = \dot{m}_A c_{PA} = (T_{A2} - T_{A1}) \tag{a}$$

$$\text{S-R A:} \quad \dot{m}_G c_{PG} = (T_{G3} - T_{G4}) = \dot{m}_F c_{PF} = (T_{F2} - T_{F1}) \tag{b}$$

Assumed temperature difference

$$T_{G3} - T_{F2} = 20\,\text{K} \tag{c}$$

Substituting data in (a), (b) and (c)

$$T_{G3} = 200 - \frac{18.75 \times 1.005}{19.75 \times 1.06}(170 - 25) = \underline{68.8\ °\text{C} = 341.95\ \text{K}}$$

From (c) $T_{F2} = 68.8 - 20 = 48.8$ °C $= \underline{321.95\ \text{K}}$

From (b) $$T_{G4} = 68.8 - \frac{1 \times 2.02}{19.75 \times 1.06}(48.8 - 25) = \underline{66.5\ °\text{C}}$$

$$= \underline{339.65\ \text{K}}$$

Adiabatic flame temperature

Energy balance for S-R C

$$m_F(NCV) + c_{PF}(T_{F2} - T_o) + m_A c_{PA}(T_{A2} - T_o) = (T_{G1} - T_o)\sum n_k \tilde{c}_{Pk}^{h} \tag{d}$$

	N_2	O_2	CO_2	H_2O	$\sum \eta_k \tilde{c}_{Pk}^{h}$
n_k /[kmol/kg of fuel]	0.5135	0.0315	0.0725	0.065	
At 1800 °C, $\tilde{c}_P^h$ / [kJ/kmolK]	33.14	34.66	54.21	41.86	24.7
At 1860 °C, $\tilde{c}_P^h$ / [kJ/kmolK]	32.24	34.73	54.44	42.1	24.82

From (d), the energy of reactants is

$$H_{A+F} = m_F[(NCV) + \tilde{c}_{PF}(T_{F2} - T_o) + m_A \tilde{c}_{PA}(T_{A2} - T_o)$$

$$= 45294 \text{ kJ/kmol of fuel} \qquad \text{(e)}$$

Hence, using the trial temperature in the above table T_{G1} = 1800 °C,

$$T_{G1} = T_o + \frac{H_{A+F}}{\sum \eta_k \tilde{c}_{Pk}^{h}} = 25 + \frac{45294}{24.7} = 1859\,°C$$

Using the second trial value of $\sum_k \eta_k \tilde{c}_{Pk}^{h}$ corresponding to T_{G1} = 1860 °C, we get

$$T_{G1} = 25 + \frac{45294}{24.82} = \underline{1850\ °C}$$

which is close enough to the trial value of 1860 °C.

<u>Chemical exergy of the fuel oil</u>

From (C.6) in the Book,

$$\varphi = 1.0401 + 0.1728\frac{h}{c} \qquad \text{(f)}$$

Since c = 0.87 and h = 0.13

$$\varphi = 1.066$$

Hence

= <u>45302 kJ/kg</u>

<u>Irreversibililties</u>:

<u>Calculation of changes of exergy of preheater streams using the "across the unit" method</u>

This method is convenient and applicable when it is possible to assume P = const and c_P = const. In general,

$$\dot{H}_2 - \dot{H}_1 = \Delta\dot{H}_{1,2} = \dot{m}\int_1^2 c_P \mathrm{d}T \qquad \text{(g)}$$

and

$$\dot{S}_2 - \dot{S}_1 = \Delta\dot{S}_{1,2} = \dot{m}\int_1^2 \frac{c_P \mathrm{d}T}{T} \tag{h}$$

Case (i) $\quad T \neq \text{const}$

$$\frac{\Delta\dot{S}_{1,2}}{\Delta\dot{H}_{1,2}} = \frac{\int_1^2 \mathrm{d}T/T}{\mathrm{d}T} = \frac{\ln T_2/T_1}{T_2 - T_1} \tag{i}$$

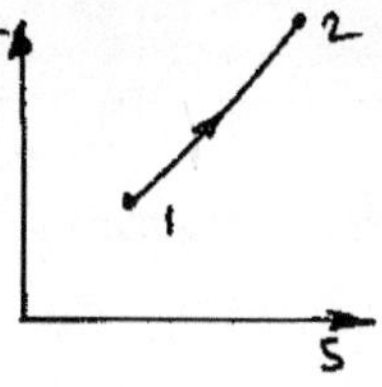

Fig.5.5.c

Exergy rate change for a stream

$$\dot{E}_2 - \dot{E}_1 = \Delta H_{1,2} - T_o \Delta\dot{S}_{1,2}$$

Using (i)

$$\dot{E}_2 - \dot{E}_1 = \Delta H_{1,2}\left(1 - \frac{T_o}{T_2 - T_1}\ln\frac{T_2}{T_1}\right) \tag{j}$$

Case (ii) $\quad T = \text{const} = T_c$ (not relevant here!)

As will be seen from the diagram

$$\frac{\Delta\dot{S}_{1,2}}{\Delta\dot{H}_{1,2}} = \frac{1}{T_C}$$

Hence

$$\dot{E}_2 - \dot{E}_1 = \Delta H_{1,2}\left[1 - \frac{T_o}{T_C}\right] \tag{k}$$

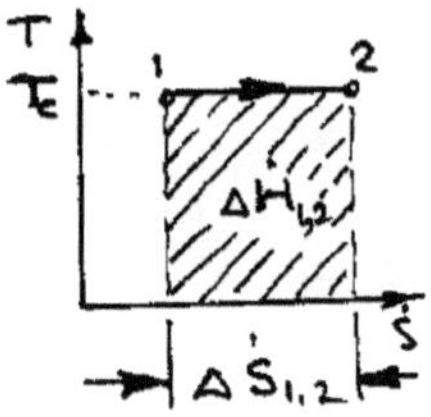

Fig.5.5.d

Subregion A - Fuel preheater

For the gas stream, using (j)

$$E_{G3} - E_{G4} = m_G c_{PG}(T_{G3} - T_{G4})\left[1 - \frac{T_o}{T_{G3} - T_{G4}}\ln\frac{T_{G3}}{T_{G4}}\right] \tag{k}$$

$$= 19.75 \times 1.06(68.8 - 66.5)\left[1 - \frac{298.15}{68.8 - 66.5}\ln\frac{341.95}{339.65}\right]$$

$$= 6.0 \text{ kJ}$$

For the fuel stream

$$E_{F2} - E_{F1} = 1 \times 2.02(48.8 - 25)\left[1 - \frac{198.15}{48.8 - 25}\ln\frac{321.95}{298.15}\right]$$

$$= 1.8 \text{ kJ}$$

Exergy balance for Sub region A

$$\dot{I}_A = (E_{G3} - E_{G4}) - (E_{F2} - E_{F1})$$

$$= 6.0 - 1.8 = \underline{4.2 \text{ kJ}}$$

Also, $E_{F1} = E_F^o = \underline{45302 \text{ kJ}}$

$E_{F2} = 45302 + 1.8 = \underline{45303.8 \text{ kJ}}$

Subregion B - Air preheater

For the air stream, using (j)

$$E_{A2} - E_{A1} = m_A c_{PA}(T_{A2} - T_o)\left[1 - \frac{T_o}{T_{A2} - T_o}\ln\frac{T_{A2}}{T_o}\right] \quad \text{(l)}$$

$= \underline{507.3 \text{ kJ}}$ (N.B. $E_{A1} = 0 \therefore \dot{E}_{A2} = \underline{507.3 \text{ kJ}}$)

For the products stream

$$E_{G2} - E_{G3} = m_G c_{PG}(T_{G1} - T_{G2})\left[1 - \frac{T_o}{T_{G1} - T_{G2}}\ln\frac{T_{G1}}{T_{G2}}\right] \quad \text{(m)}$$

$= \underline{719.7 \text{ kJ}}$

Exergy balance for Subregion B

$$\dot{I}_B = (E_{G2} - E_{G3}) - (E_{A2} - E_{A1}) \quad \text{(n)}$$

$= 719.7 - 507.3 = \underline{212.4 \text{ kJ}}$

Subregion C - Adiabatic combustion

From the exergy balance

$$\dot{I}_C = E_{F2} + E_{A2} - E_{G1} \quad \text{(o)}$$

To calculate E_{G1} we use

$$E_{G1} = n_G\left[(T_{G1} - T_o)\sum_k x_k \tilde{c}_{P,k}^{\varepsilon} + \tilde{R}T_o \ln\frac{P_{G1}}{P_o}\right] \quad \text{(p)}$$

From Table D3

	N_2	O_2	CO_2	H_2O	$\sum x_k \tilde{c}_{P,k}^{\varepsilon}$
x_k	0.752	0.0462	0.106	0.095	
$\tilde{c}_P^{\varepsilon}$/[kJ/kmolK] at 1850 °C	22.94	23.99	38.06	29.20	25.17

With $n_G = 0.6825$ kmol/kg of fuel

$E_{G1} = 0.6825(1850 - 25)25.17 = \underline{31348 \text{ kJ}}$

Irreversibility of the combustion process

$\dot{I}_C = 45303.8 + 507.3 - 31348 = \underline{14463.1 \text{ kJ}}$

Subregion D - Steam generator

Steam properties at: $P_s = 150$ bar

$T_{s1} = 25$ °C $\quad$ $T_{s2} = 500$ °C

$h_{s1} = 104.8$ kJ/kg $\quad$ $h_{s2} = 3309$ kJ/kg

$s_{s1} = 0.367$ kJ/kgK $\quad$ $s_{s2} = 6.345$ kJ/kgK

Energy balance for S-R D

$$H_{G1} - H_{G2} = m_s(h_{s2} - h_{s1}) \tag{r}$$

Since $H_{G1} = H_{A+F}$, we have from (e)

$H_{G1} = 45294$ kJ

$H_{G2} = m_G c_{PG}(T_{G2} - T_o)$

$= 19.75 \times 1.06\ (200 - 25) = 3663.6$ kJ

Mass flow rate of steam/water from (r)

$$m_s = \frac{45294 - 3663.6}{3309 - 104.8} = 13.0 \text{ kg/kg of fuel}$$

Exergy balance for S-R D

$$I_D = (E_{G1} - E_{G2}) - (E_{s2} - E_{s1}) \tag{s}$$

For the H_2O stream

$$E_{s2} - E_{s1} = m_s - [(h_{s2} - h_{s1}) - T_o(s_{s2} - s_{s1})] \tag{t}$$

$= \underline{18484 \text{ kJ}}$

As shown in (p) $\quad E_{G2} = n_G(T_{G1} - T_o)\sum_k x_k \tilde{c}^{\varepsilon}_{P.k}$

	N_2	O_2	CO_2	H_2O	$\sum x_k \tilde{c}^{\varepsilon}_{P,k}$
X_k	0.752	0.0462	0.106	0.095	
$\tilde{c}^{\varepsilon}_P$/[kJ/kmolK] at 1850 °C	6.34	6.45	9.09	7.22	6.715

Hence, $\quad E_{G2} = 0.6825\ (100 - 25)\ 6.715 = \underline{802 \text{ kJ}}$

Irreversibility of the heat transfer process

$I_D = 31348 - 802 - 18484 = \underline{12062 \text{ kJ}}$

Taking $E_{s1} \cong 0$ (since $T_{s1} = T_o$)

We have from (t)

$E_{s2} = \underline{18484 \text{ kJ}}$

<u>Subregion E - Mixing of products with atmospheric air</u>

From the exergy balance

$$I_{\mathrm{E}} = E_{\mathrm{G4}}$$

From (k) and (m) $\quad E_{\mathrm{G4}} = E_{\mathrm{G2}} - [6.0 + 719.7]$ kJ

With $E_{\mathrm{G2}} = 802$ kJ $\quad E_{\mathrm{G4}} = \underline{76.3 \text{ kJ}}$

$\therefore \underline{I_{\mathrm{E}} = 76.3 \text{ kJ}}$

Rational efficiency of the plant

$$\psi = \frac{E_{\mathrm{s2}}}{E_{\mathrm{F1}}} = \frac{18484}{45302} = \underline{0.41}$$

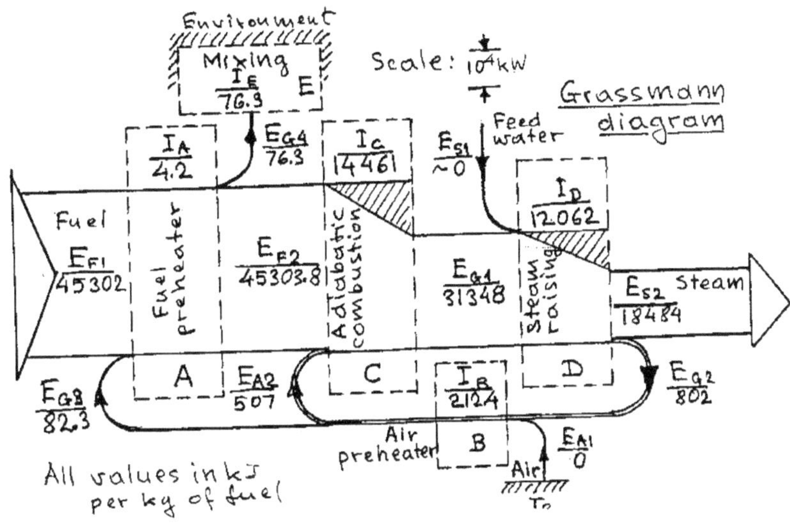

Fig. 5.5.e

Problem 6.1

To determine: The number of years needed to repay a loan of £12 × 10^6 with 15% annual interest, compounded monthly, using monthly repayments of £180,000

Analysis: Monthly repayment

$$R = Ca^{\mathrm{c}} = C\frac{i(1+i)^{\mathrm{n}}}{(1+i)-1} = £180,000 \qquad \text{(a)}$$

where,

C = £12 × 10^6 - the loan

a^{c} = capital - recovery factor

i = interest rate, monthly

n = number of months of repayment

From (a)

$$a^{c} = \frac{180000}{12\times10^{6}} = 0.015\,(\text{month})^{-1}$$

Monthly interest rate

$$i = \frac{0.15}{12} = 0.0125$$

Denoting by $X = (1 + i)^{n}$ we get from (a)

$$X = \frac{a^{c}}{a^{c} - i} = \frac{0.015}{0.015 - 0.0125} = 6$$

$\therefore\;\; 6 = (1.0125)^{n}$

$$n = \frac{\ln 6}{\ln 1.0125} = \underline{144.2 \text{ months}}$$

$$\cong \underline{12 \text{ years}}$$

<u>Problem 6.2</u>

<u>The plant</u>: A steam boiler.

<u>To determine</u>: The minimum cost of exergy of the steam delivered.

Given data:

$$\dot{E}_{S} = 2\times10^{4}\text{ kW}$$

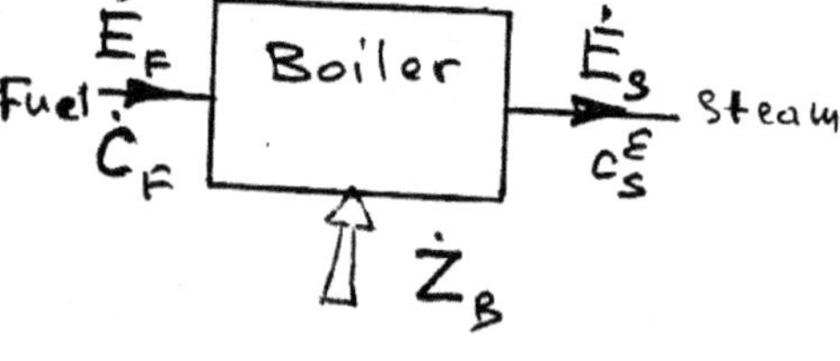

Fig. 6.2

Fuel cost rate

$$\dot{C}_{F} = 500\,£/\text{h}$$

Capital cost of the boiler

$$\dot{C}_{B} = £4.5\times10^{6}$$

Period of operation per year $\quad t_{op} = 7000$ h

Capital recovery factor $\quad a^{c} = 0.1868\ \text{y}^{-1}$

<u>Analysis</u>:

Unit cost of exergy of steam, from the cost equation

$$c_{S}^{\varepsilon} = \frac{\dot{C}_{F} + \dot{Z}_{B}}{\dot{E}_{S}} \tag{a}$$

where,

$$\dot{C}_{F} = c_{F}^{\varepsilon}\dot{E}_{F} \tag{b}$$

$$\dot{Z}_{B} = \frac{a^{c}}{t_{op}}C_{B}^{c} \tag{c}$$

Substituting numerical values

$$\dot{C}_F = \frac{500\pounds/h}{3600\ s/h} = 0.139\ \pounds/s$$

$$\dot{Z}_B = \frac{0.1868[y]^{-1} \times \pounds 4.5 \times 10^6}{7000h/y \times 3600\ s/h} = 0.0333\ \frac{\pounds}{s}$$

From (a)

$$\dot{C}_S^{\varepsilon} = \frac{(0.139 + 0.0333)\ \pounds/s}{2 \times 10^4\ kW} = \underline{8.61 \times 10^{-6}\ \frac{\pounds}{kJ}}$$

$$= \underline{3.1\ p/kWh}$$

Comment: Usually other costs, such as maintenance, also have to be considered.

Problem 6.3

The plant: An ammonia vapour-compression refrigerator described in Problem 3.12.

To determine: (i) Irreversibility rates in the four main components,

(ii) changes in the irreversibility rates resulting from a change in the evaporator pressure as fractions of the change in the total plant irreversibility,

(iii) coefficient of structural bonds for the evaporator.

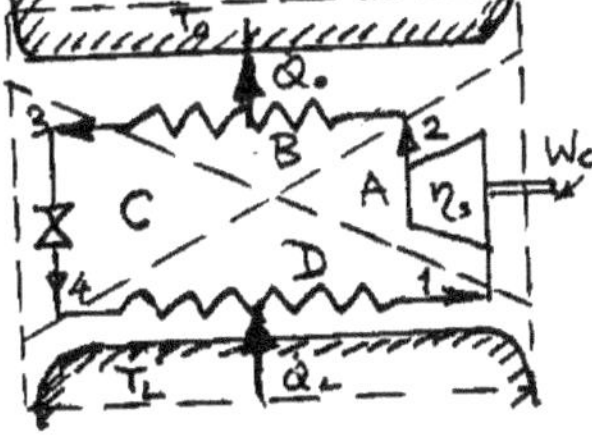

Fig. 6.3.a

Given data:

$$\eta_s = 0.73 \qquad \dot{Q}_L = 100\ kW$$

New $P_{EVAP} = 2.680$ bar

From NH_3 tables:

$h_1 = 1430.5$ kJ/kg $\qquad h_3 = h_4 = 323.1$ kJ/kg

$s_1 = 5.504$ kJ/kgK $\qquad s_3 = 1.204$ kJ/kg

$s_2^1 = s_1$ hence by interpolation

$h_{2'} = 1643.2$ kJ/kg

$$h_2 = h_1 + \frac{h_{2'} - h_1}{T_{EVAP}} = 5.504 - \frac{1430.5 - 323.1}{261.15} = 1.2635 \frac{kJ}{kgK}$$

Assumptions: $\dot{Q}_L$ is the same both when $P_{EVAP} = 2.908$ bar (Problem 3.12) and when it is 2.680 bar as in this Problem. Stray heat transfer and pressure losses as well as ΔKE and ΔPE are negligible.

Analysis:

Mass flow rate of NH_3

$$\dot{m} = \frac{\dot{Q}_L}{h_1 - h_4} = \frac{100}{1430.5 - 323.1} = 0.0903 \text{ kg/s}$$

Irreversibility rates (from G-S relation):

Subregion A - compressor

$$\dot{I}_A = \dot{m}T_o(s_2 - s_1) = \underline{5.616 \text{ kW}}$$

Subregion B - condenser

$$\dot{I}_B = \dot{m}T_o(s_3 - s_2) - \dot{Q}_o$$

but

$$\dot{Q}_o = \dot{m}(h_2 - h_3) = \underline{126.3 \text{ kW}}$$

Hence, $\dot{I}_B = \underline{4.915 \text{ kW}}$

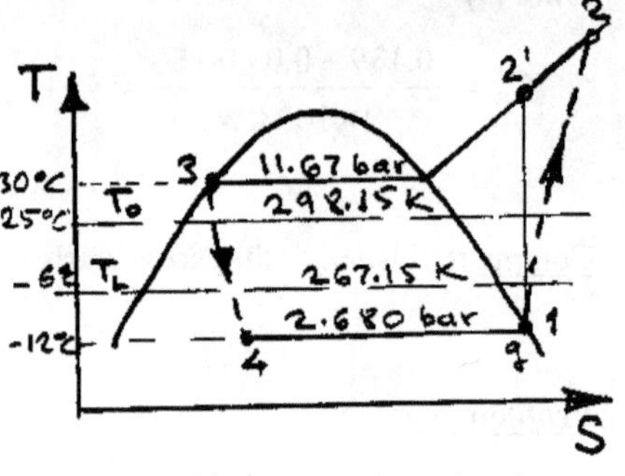

Fig. 6.3.b

Subregion C - throttling valve

$$\dot{I}_C = \dot{m}T_o(s_4 - s_3) = \underline{1.602 \text{ kW}}$$

Subregion D - evaporator

$$\dot{I}_D = T_o\left[\dot{m}(s_1 - s_4) - \frac{\dot{Q}_L}{T_L}\right] = \underline{2.563 \text{ kW}}$$

Changes in the irreversibility rates

For the case when $P_{EVAP} = 2.908$ bar (see Problem 3.12), with $\dot{Q}_L = 100 \text{ kW}$

$$\dot{m}_1 = \frac{100}{1433.0 - 323.1} = \underline{0.0901 \text{ kg/s}}$$

Using this value of $\dot{m}_1$ and specific irreversibility rates calculated earlier:

Subregion	A	B	C	D	Total
$\dot{I}_1$/kW; $P_{EVAP} = 2908$ bar	5.265	4.604	1.424	1.702	12.99
$\dot{I}_2$/kW; $P_{EVAP} = 2680$ bar	5.616	4.915	1.602	2.563	14.696
$\Delta\dot{I} = (\dot{I}_2 - \dot{I}_1)/\text{kW}$	0.351	0.311	0.178	0.861	1.706
$\Delta\dot{I}_k / \Delta\dot{I}_{TOT}$	0.207	0.182	0.104	0.506	1

Coefficient of structural bonds

$$\sigma_{\mathrm{D,PEV}} = \left(\frac{\Delta I_{\mathrm{TOT}}}{\Delta I_{\mathrm{EV}}}\right)_{p_{\mathrm{EV}}} = \mathrm{var}.$$

$$= \frac{1}{0.506} = \underline{1.98}$$

Comment: The reduction in the evaporator pressure is accompanied by an increase in the temperature difference in heat transfer from 4K to 6K. This results in an increase in irreversibility rates in all the four components, the largest being in the evaporator itself. This change decreases in the different components in the direction of flow of NH_3. The coefficient of structural bonds is a measure of how much a change in a local irreversibility affects the irreversibility of the whole plant. Here, it is found that, because of the interaction between the components resulting from the plant structure, the change in the irreversibility of the whole plant is almost twice that of the evaporator.

Problem 6.4

The plant: An ammonia vapour-compression refrigerator described in Problem 3.12.

To determine: The optimum temperature difference for heat transfer in the evaporator and the corresponding irreversibility rate, heat transfer area and the capital cost of the heat exchanger.

Given data

Capital cost - empirical expression:

$$C_{\mathrm{D}}^{\mathrm{C}} = K_{\mathrm{B}} + \frac{K_{\mathrm{A}}\dot{Q}_{\mathrm{D}}}{U\Delta T} \qquad \text{(a)}$$

Irreversibility rate for heat transfer at constant temperatures T_1 and T_2

$$\dot{I}_{\mathrm{D}} = \dot{Q}_{\mathrm{D}}\frac{T_{\mathrm{o}}\Delta T}{T_{\mathrm{M}}^{2}} \qquad \text{(b)}$$

where (see Problem 3.20)

$$\Delta T = T_{\mathrm{L}} - T_{\mathrm{EVAP}} \qquad \text{(c)}$$

$$T_{\mathrm{M}} = \sqrt{T_{\mathrm{L}}T_{\mathrm{EVAP}}} \qquad \text{(d)}$$

$K_{\mathrm{A}} = 200$ £/m^2, $K_{\mathrm{B}} =$ £460, $U = 2$ kW/m^{2}K,

$\dot{Q}_{\mathrm{D}} = 100$ kW, $T_{\mathrm{o}} = 288$ K, $T_{\mathrm{M}} \cong 266$ K

$t_{\mathrm{op}} = 4000$ h/y, $a^{\mathrm{c}} = 0.1868$ y^{-1}, $T_{\mathrm{L}} = 267.15$ K

$C_{\mathrm{IN}}^{\varepsilon} = 0.05$ £/kWh, $\sigma_{\mathrm{D},\Delta T} = 1.98$ (from Problem 6.3)

Assumptions: As in Problem 3.12

Analysis:

Using expressions (6.58 and (6.61) with $x_i = \Delta T$, the constant temperature difference, we have

$$\frac{\partial C_T}{\partial(\Delta T)} = t_{op} C_D^I \frac{\partial \dot{I}_E}{\partial(\Delta T)} + a^c \frac{\partial C_D^c}{\partial(\Delta T)} \quad \text{(e)}$$

where

$$C_D^I = c_{IN}^{\varepsilon} \sigma_{D,\Delta T} \quad \text{(f)}$$

Substituting for C_D^C and $\dot{I}_D$ from (a) and (b)

$$\frac{\partial C_T}{\partial(\Delta T)} = t_{op} c_{IN}^{\varepsilon} \sigma_{D,\Delta T} \frac{\partial}{\partial(\Delta T)}\left[\frac{T_o \Delta T}{T_M^2}\right] + a^c \frac{\partial}{\partial(\Delta T)}\left[\frac{k_A \dot{Q}_D}{U(\Delta T)}\right] \quad \text{(g)}$$

Differentiating and equating (g) to zero

$$\frac{1}{T_M^2}\left(t_{op} c_{IN}^{\varepsilon} \sigma_{D,\Delta T} T_o \dot{Q}_D\right) = \frac{a^c k_A \dot{Q}_D}{(\Delta T)_{opt}^2 U} \quad \text{(h)}$$

Hence,

$$(\Delta T)_{opt} = T_M \sqrt{\frac{a^c k_A}{U t_{op} c_{IN}^{\varepsilon} \sigma_{D,\Delta T} T_o}} \quad \text{(i)}$$

Substituting numerical data

$$(\Delta T)_{opt} = 266 \sqrt{\frac{m^2 k \times 0.1868 \mathrm{y}^{-1} \times 200 £ \times \mathrm{kwh}}{2\,\mathrm{kW} \times 4000\,\mathrm{hy}^{-1} \times m^2 \times 0.05 £ \times 1.98 \times 300\,\mathrm{K}}}$$

$= \underline{3.32\ \mathrm{K}}$ - first approximation

From (c)

$$T_{EVAP} = T_L - (\Delta T)_{opt} = 267.15\text{-}3.32\text{=}263.82\ \mathrm{K}$$

As this is very close to the assumed value of 266 K, no further iteration is required.

Irreversibility rate, from (b)

$$\left(\dot{I}_D\right)_{opt} = 100 \frac{288 \times 3.32}{266^2} = \underline{1.35\ \mathrm{kW}}$$

Heat transfer area

$$\left(A_D\right)_{opt} = \frac{\dot{Q}_D}{U(\Delta T)_{opt}} = \frac{100}{2 \times 3.32} = \underline{15.06\ \mathrm{m}^2}$$

Capital cost of the heat exchanger, from (a)

$$C_{\mathrm{D}}^{\mathrm{c}} = £460 + \frac{200[£/\mathrm{m}^2]\times 100\,\mathrm{kW}}{2[\mathrm{kW/m^2K}]\times 3.32\,\mathrm{K}} = \underline{£3472}$$

Problem 6.5

The plant: Coal-fired, steam, electric power plant described in Problem 5.1

To determine: Using the data given in Problem 5.1 and a new turbine isentropic efficiency of 0.75 calculate:

(i) irreversibility rates of the main sub regions of the plant,
(ii) changes in these irreversibilities as a percentage of that of the plant as a whole,
(iii) coefficient of structural bonds for the turbine.

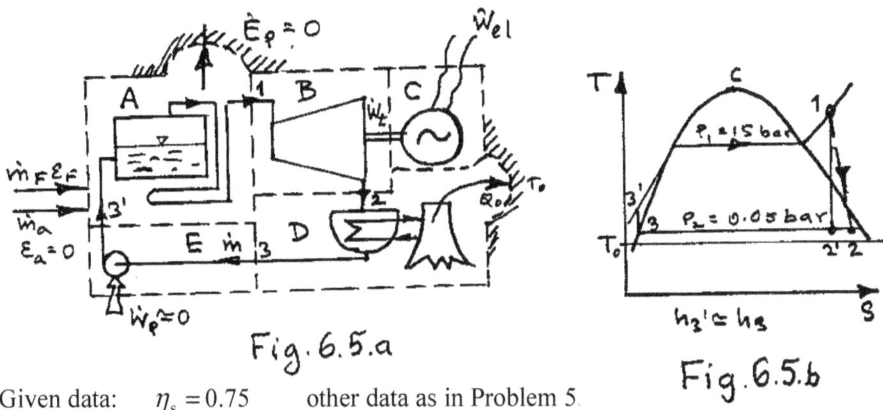

Fig. 6.5.a

Fig. 6.5.b

Given data: $\eta_s = 0.75$ other data as in Problem 5

Assumptions: The power output is the same as in Problem 5.1

Analysis:

Steam properties (from tables) for the cycle:

$h_1 = 3039$ kJ/kg $\quad s_1 = 6.919$ kJ/kgK

$h_2 = 2342.0$ kJ/kg $\quad s_2 = 7.678$ kJ/kgK

$h_3 = 138$ kJ/kgK $\quad s_3 = 0.476$ kJ/kgK

Using expressions given in Problem 5.1, the new values of specific irreversibilities of the sub regions have been calculated:

$i_{\mathrm{A}} = 2767.2\,\mathrm{kJ/kg}$

$i_{\mathrm{B}} = 214.9\,\mathrm{kJ/kg}$

$i_{\mathrm{C}} = 69.7\,\mathrm{kJ/kg}$

$\mathrm{i}_{\mathrm{D}} = 165.3\,\mathrm{kJ/kg}$

$i_{\mathrm{E}} = 0$

Mass flow rate of steam:

Turbine power output $\dot{W}_t = \dfrac{\dot{W}_{el}}{\eta_{el}} = 5555.6\ \text{kW}$

$\dot{m}_s = \dot{W}_t / (h_1 - h_2) = 7.97\ \text{kg/s}$

Irreversibility rates

$\dot{I}_k = \dot{m}_s i_k$

Comparison with results obtained in Problem 5.

Subregion	A	B	C	D	E	Total
$\dot{I}_k$/kW; ($\eta_s = 0.75$)	22054.6	1712.8	555.5	1317.4	0	25640
$\dot{I}_k^{'}$/kW; ($\eta_s = 0.70$)	23604	2200	555.5	1430	0	27789
$(\dot{I}_k^{'} - \dot{I}_k)$/kW	1549.4	487.2	0	112.6	0	2149
$(\dot{I}_k^{'} - \dot{I}_k)/\Delta\dot{I}_{TOT}$	0.72	0.227	0	0.052	0	1

Coefficient of structural bonds for the turbine due to a change in η_s

$$\sigma_{t,\eta_s} = \left(\frac{\Delta\dot{I}_{TOT}}{\Delta\dot{I}_t}\right)_{\eta_s = \text{var}} = \frac{2149}{487.2} = \underline{4.41}$$

Comment: As shown in the above table, the largest relative reduction in irreversibility rate (0.72) associated with the increase in η_s takes place in the boiler (Subregion A). This is achieved purely through a reduction in the steam demand necessary to deliver the required power output, without any improvement in boiler efficiency (note that there is no change in i_A). Since the boiler irreversibility rate is by far the largest of all the plant elements, the reduction is also the largest. Although the improvement in performance, by increasing η_s, has been carried out on the turbine (Sub region B), the effect of it on irreversibility rate is 3.47 times larger outside this sub region than inside it. This leads to the high value of the coefficient of structural bonds which has been obtained.

Problem 6.6

The plant: Coal-fired, steam, electric power plant described in Problem 5.1 and considered in Problem 6.5

To determine: Using the structural method, calculate the optimum isentropic efficiency and the capital cost of the turbine.

Given data:

Empirical expression for the capital cost of the turbine:

$$C_{\mathrm{t}}^{\mathrm{c}} / £ = -2.44 \times 10^{6} + 10.15 \times 10^{6} \eta_{\mathrm{s}}^{1.8} \quad \text{(a)}$$

$$\text{for } 0.6 < \eta_{\mathrm{s}} < 0.9 \quad \text{(b)}$$

$a^{\mathrm{c}} = 0.124\ \mathrm{y}^{-1}$ $\qquad t_{\mathrm{op}} = 6000\ \mathrm{h/y}$

$NCV = 25.0\ \mathrm{MJ/kg}$ $\qquad c_{\mathrm{F}}^{\mathrm{sp}} = £65 / \mathrm{tonne}$

$\sigma_{\mathrm{t},\eta_{\mathrm{s}}} = 4.41$ as calculated in Problem 6.5

Assumptions: $\sigma_{\mathrm{t},\eta_{\mathrm{s}}}$ remains constant over the range of conditions specified by (b).

Analysis:

Using expressions (6.58) and (6.61) with $x_{\mathrm{i}} = \eta_{\mathrm{s}}$ we get

$$\frac{\partial C_{\mathrm{T}}}{\partial \eta_{\mathrm{s}}} = t_{\mathrm{op}} c_{\mathrm{B}}^{\mathrm{I}} \frac{\partial \dot{I}_{\mathrm{B}}}{\partial \eta_{\mathrm{s}}} + a^{\mathrm{c}} \frac{\partial C^{\mathrm{c}}}{\partial \eta_{\mathrm{s}}} \quad \text{(c)}$$

where

$$c_{\mathrm{B}}^{\mathrm{I}} \cong c_{\mathrm{IN}}^{\varepsilon} \sigma_{\mathrm{t},\eta_{\mathrm{s}}} \quad \text{(d)}$$

Equating (c) to zero

$$\left(\frac{\partial \dot{I}_{\mathrm{B}}}{\partial \eta_{\mathrm{s}}} \right)_{\mathrm{opt}} = - \frac{a^{\mathrm{c}}}{t_{\mathrm{op}} C_{\mathrm{IN}}^{\varepsilon} \sigma_{\mathrm{t},\eta_{\mathrm{s}}}} \frac{\partial C^{\mathrm{c}}}{\partial \eta_{\mathrm{s}}} \quad \text{(e)}$$

As shown in Problem 4.9, the irreversibility of a steam turbine can be expressed as

$$\dot{I} = \dot{m}_{\mathrm{s}} (h_1 - h_2) \left[\frac{1}{\eta_{\mathrm{s}}} - 1 \right] \frac{T_{\mathrm{o}}}{T_{\mathrm{c}}} \quad \text{(f)}$$

Hence we get

$$\frac{\partial \dot{I}_{\mathrm{B}}}{\partial \eta_{\mathrm{s}}} = -\dot{W}_{\mathrm{t}} \frac{T_{\mathrm{o}}}{T_{\mathrm{c}}} \frac{1}{\eta_{\mathrm{s}}^{2}} \quad \text{(g)}$$

where

$$\dot{W}_{\mathrm{t}} = \dot{m}_{\mathrm{s}} (h_1 - h_2) \quad \text{(h)}$$

Differentiating (a) with respect to η_{s}

$$\frac{\partial C_{\mathrm{t}}^{\mathrm{c}}}{\partial \eta_{\mathrm{s}}} = K_{\eta_{\mathrm{s}}}^{0.8} \quad \text{(i)}$$

where $\qquad K = £18.27 \times 10^{6}$

Substituting (g) and (i) in (e)

$$\frac{1}{\eta_s^2}\frac{T_o}{T_c}\dot{W}_t = \frac{a^c K}{t_{op} C_{IN}^{\varepsilon} \sigma_{t,\eta_s}} \eta_s^{0.8} \tag{j}$$

Hence,

$$\eta_s = \left[\frac{\dot{W}_t t_{op} c_{IN}^{\varepsilon} \sigma_{t,\eta_s} T_o}{a^c K T_c}\right]^{\frac{1}{2.8}} \tag{k}$$

Unit price of exergy of coal, c_{IN}^{ε}

Exergy of coal $\qquad \varepsilon_F = \varphi(NCV) = 1.06 \times 25\,\text{MJ/kg}$

$= 26.5$ MJ/kg

Specific cost of coal

$c_F^{sp} = £65/\text{tonne}$

Hence,

$$c_{IN}^{\varepsilon} = \frac{c_F^{sp}}{\varepsilon_F} = \frac{65£ \ \text{tonne} \ \text{kg}}{\text{tonne} \times 10^3 \ \text{kg} \times 26.5 \times 10^3 \ \text{kJ}}$$

$= \underline{2.45 \times 10^{-6} \ £/\text{kJ}}$

Optimum isentropic efficiency

$$\eta_s = \left[\frac{5555.6[\text{kJ/s}]600[Wy]3600\left[\frac{\text{s}}{\text{h}}\right]2.45\times10^{-6}\left[\frac{£}{\text{kJ}}\right]4.41\times283.15\,\text{K}}{18.27\times10^{-6}£\times0.124\,\text{y}^{-1}\times308\text{K}}\right]^{\frac{1}{2.8}}$$

$= \underline{0.80}$

Optimum capital cost of the turbine

Using (a) with $\eta_s = 0.8$

$C_t^c / £ = -2.44\times10^{-6} + 10.15\times10^6 \times 0.8^{1.8}$

$= \underline{4.35 \times 10^6}$

Comment: Increasing the value of η_s from 0.7 to the optimum value of 0.8 improves the rational efficiency of the plant from 0.152 (see Problem 5.1) to (as can be calculated) 0.174, whilst the capital cost of turbine increases from £2.9 × 10^6 to £4.35 × 10^6.

Problem 6.7

The plant: Dual purpose, refrigerator - heat pump plant as described in Problem 5.2.

To determine: Unit costs of exergy of the two forms of output and the unit cost of energy delivered to the swimming pool.

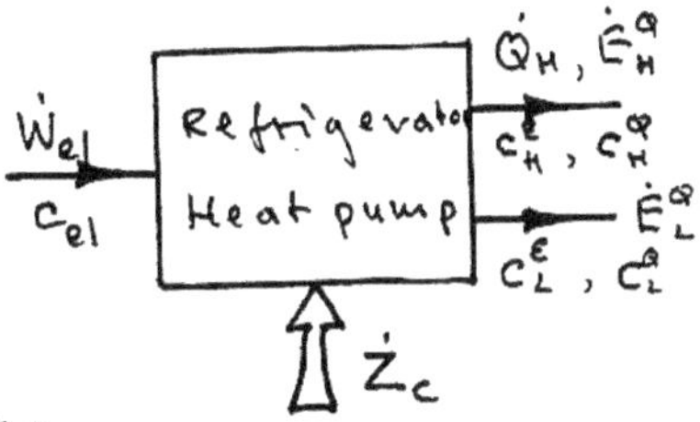

Given data: $C^c = £64000$

$i = 0.12\ \text{y}^{-1}$

$N_y = 20\text{y}, \qquad t_{op} = 7800\ \text{h/y}$

$c_{el} = 0.07\ £/\text{kWh} = 19.4 \times 10^{-6}\ £/\text{kJ}$

Fig.6.7.

Other data as in Problem 5.2

Assumptions: Both forms of output are of equal importance; period of repayment is the same as the projected life of the plant, i.e. $c_H^{\varepsilon} = c_L^{\varepsilon} = c^{\varepsilon}$.

Analysis: Interest rate per month $\quad i = 0.12/12$

$$= 0.01\ \text{month}^{-1}$$

Capital recovery factor

$$a^c = \frac{0.01(1+0.01)^{240}}{(1.001)^{240} - 1} = 0.011\ \text{month}^{-1}$$

Capital investment rate

$$\dot{Z}_c = \frac{a^c C^c}{t_{op}} = \frac{0.011\,\text{month}^{-1} \times 64000£ \times 12\,\text{month y}^{-1}}{7800\text{h y}^{-1} \times 3600\,\text{s h}^{-1}}$$

$$= 0.30 \times 10^{-3}\ £/\text{s}$$

From the money balance

$$C^{\varepsilon} = \frac{\dot{W}_{el} c_{el} \dot{Z}_c}{\dot{E}_H^Q + \dot{E}_L^Q} = \frac{15.29\frac{\text{kJ}}{\text{s}} 19.4 \times 10^{-6}\frac{£}{\text{kJ}} + 0.30 \times 10^{-3}\frac{£}{\text{s}}}{(1.69 + 4.46)\frac{\text{kJ}}{\text{s}}} = \underline{97 \times 10^{-6}\ \frac{£}{\text{kJ}}}$$

$$= \underline{0.35\ \frac{£}{\text{kWh}}}$$

Unit cost of energy delivered to the pool

Energy cost rate = Exergy cost rate

i.e. $\quad \dot{Q}_H \times c_H^Q = \dot{E}_H^Q \times c_H^{\varepsilon}$ now, since $\dot{E}_H^Q = \dot{Q}_H \dfrac{T_H - T_o}{T_H}$

$$\therefore \qquad c_{\rm H}^{\rm Q} = \frac{T_{\rm H} - T_o}{T_{\rm H}} c_{\rm H}^{\varepsilon}$$

$$= \frac{301 - 293}{301} \times 97 \times 10^{-6} \frac{\pounds}{\rm kJ} = \underline{2.58 \times 10^{-6}\ \pounds/\rm kJ}$$

$$= \underline{0.93\rm p/kWh}$$

Comment: The unit cost of exergy, c^{ε}, delivered by the plant as heating and refrigeration is five times larger than the unit cost of the input exergy, $c_{\rm el}$. As follows from the expression

$$c^{\varepsilon} = \frac{c_{\rm el}}{\psi} + \frac{\dot{Z}_{\rm c}}{\dot{E}_{\rm H}^{\rm Q} + \dot{E}_{\rm L}^{\rm Q}}$$

where

$$\psi = \frac{\dot{E}_{\rm H}^{\rm Q} + \dot{E}_{\rm L}^{\rm Q}}{\dot{W}_{\rm el}} = 0.402$$

the magnitude of c^{ε} is partly dependent on ψ whose contribution accounts for about half of c^{ε}. The other half is due to the contribution from the investment cost rate per unit of the output exergy.

When evaluated per unit of the heating delivered to the swimming pool, the unit cost is comparatively low.

Problem 6.8

The plant: Gas-fired water heater delivering 60 kW to the swimming pool.

To determine: (i) The rational efficiency of the plant,
(ii) unit costs of both energy and exergy delivered to the swimming pool.

Given data:

For the fuel gas:

$NCV = 50014$ kJ/kg

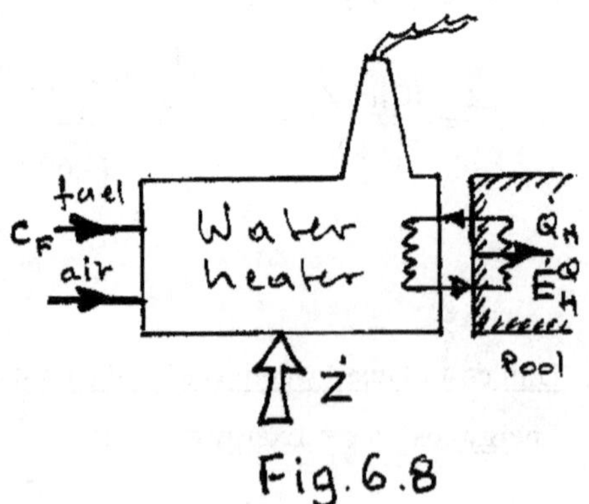

Fig.6.8

$\varepsilon_{\rm F} = 52145$ kJ/kg

$c_{\rm F} = 0.014$ £/kWh

$\eta = 0.78$

Also, data from Problems 5.2 and 6.7

Assumptions: The capital cost includes the cost of installation. The loan is arranged on the same terms as in the case of the refrigerator-heat pump plant in Problem 6.7.

Analysis:

Unit cost of the fuel gas

$$c_F = 0.014 \text{ £/kWh} = 3.89 \times 10^{-6} \text{ £/kJ}$$

Capital investment rate

$$\dot{Z}_c = \frac{a^c C^c}{t_{op}} = \frac{0.011 \text{ month}^{-1} \times 6000\text{£} \times 12 \text{ month/y}}{7800 \text{ h/y} \times 3600 \text{ s/h}}$$

$$= 28.2 \times 10^{-6} \text{ £/s}$$

Enthalpy input rate with the fuel

$$\dot{H}_F = \frac{\dot{Q}_H}{\eta} = \frac{60 \text{ kW}}{0.78} = 76.9 \text{ kW}$$

Cost equation in terms of energy unit costs

$$c_F \dot{H}_F + \dot{Z}_c = \dot{Q}_H c_H^Q$$

Hence,

$$c_H^Q = \frac{c_F \dot{H}_F + \dot{Z}_c}{\dot{Q}_H}$$

$$= \frac{3.89 \times 10^{-6} \dfrac{\text{£}}{\text{kJ}} \times 76.9 \dfrac{\text{kJ}}{\text{s}} + 28.2 \times 10^{-6} \dfrac{\text{£}}{s}}{60 \dfrac{\text{kJ}}{\text{s}}}$$

$$= \underline{5.46 \times 10^{-6} \text{ £/kJ}} = \underline{0.0196 \text{ £/kWh}}$$

Unit cost of the thermal exergy output

$$\dot{Q}_H c_H^Q = \dot{E}_H^Q c_H^\varepsilon$$

$$\therefore \quad c_H^\varepsilon = \frac{T_H}{T_H - T_o} c_H^Q = \frac{301}{301 - 293} \times 5.46 \times 10^{-6} \frac{\text{£}}{\text{kJ}}$$

$$= \underline{0.205 \times 10^{-3} \text{ £/kJ}} = \underline{0.738 \text{ £/kWh}}$$

Rational efficiency of the plant

$$\psi = \frac{\text{exergy output}}{\text{exergy input}} = \frac{E_H'^Q}{\dot{E}_F} = \frac{\dot{Q}_H (T_H - T_o)}{\dot{m}_F \varepsilon_F T_H}$$

Mass flow rate of fuel gas

$$\dot{m}_F = \frac{\dot{H}_F}{(NCV)} = \frac{76.9\frac{\text{kJ}}{\text{s}}}{50014\frac{\text{kJ}}{\text{kg}}} = 1.54\times10^{-3}\ \text{kg/s}$$

Substituting numerical values

$$\psi = \frac{60\frac{\text{kJ}}{\text{s}}(301-293)\ \text{K}}{1.54\times10^{-3}\frac{\text{kg}}{\text{s}}\times52145\frac{\text{kJ}}{\text{kg}}\times301\ \text{K}} = \underline{0.02}$$

<u>Comment:</u>

Plant	ψ	$\frac{c_H^Q}{\text{p/kWh}}$	$\frac{c_H^\varepsilon}{\text{p/kWh}}$
Refrigerator-heat pump	0.402	0.93	35
Water heater	0.02	1.96	73.8

As will be seen from the above table, the refrigerator-heat pump plant has both higher rational efficiency and is capable of delivering heating of the pool at about half the cost of that of the water heater.

Problem 6.9

<u>The plants</u>: A series of vapour-compression plants of the same type and the same structure.

<u>To determine</u>: Using a method of approximate thermo-economic modelling

(i) calculate optimum value of rational efficiency ψ and the corresponding capital cost of C^c, for a plant of given parameters, and
(ii) investigate the effect of variation in i and c_{IN}^ε on ψ_{opt} and c^c.

Given data:

Constants for the expression:

$C_o^c = £4400$, $k = 1.850$ £/kW, m = 2.3

For (i): $\dot{Q}_{REF} = 100\ \text{kW}$, $T_c = 250$ K, $T_o = 293$

$i = 0.12\ \text{y}^{-1}$, $t_{op} = 5000$ h/y, $N_y = 20$ years

$c_{IN}^\varepsilon = 0.07$ £/kWh

For (ii) Variable parameters

$i = 0.8$ and 0.16 per year

$c_{\mathrm{IN}}^{\varepsilon} = 0.05$ and 0.09 £/kWh

<u>Analysis</u>:

Capital cost

$$C^{\mathrm{c}} = C_{\mathrm{o}}^{\mathrm{c}} + \dot{E}_{\mathrm{OUT}} k \left(\frac{\psi}{1-\psi} \right)^{\mathrm{m}} \tag{a}$$

Optimum rational efficiency

$$\psi_{\mathrm{opt}} = \frac{1}{1 + (mL)^{\frac{1}{\mathrm{m+1}}}} \tag{b}$$

where

$$L = \frac{a^{\mathrm{c}} k}{c_{\mathrm{IN}}^{\varepsilon} t_{\mathrm{op}}} \tag{c}$$

L-dimensionless number of thermo-economic similarity

(i)

Capital recovery factor

$$a^{\mathrm{c}} = \frac{i(1+i)^{\mathrm{Ny}}}{(1+i)^{\mathrm{Ny}} - 1} = \frac{0.12(1.12)^{20}}{(1.12)^{20} - 1} = 0.134 \text{ y}^{-1}$$

From (c)

$$L = \frac{0.134 \times 1850}{0.07 \times 5000} = 0.708$$

From (b)

$$\psi_{\mathrm{opt}} = \frac{1}{1 + (2.3 \times 0.708)^{\frac{1}{3.3}}} = \underline{0.463}$$

Exergy output of the refrigerator

$$\dot{E}_{\mathrm{OUT}} = -\dot{Q}_{\mathrm{REF}} \frac{T_{\mathrm{o}} - T_{\mathrm{L}}}{T_{\mathrm{L}}}$$

$$= -100 \, \frac{293 - 250}{250}$$

$$= 17.2 \text{ kW}$$

From (a)

$$C^{\mathrm{c}}_{\mathrm{opt}} = 4400 + 17.2 \times 1850\left(\frac{0.463}{1-0.463}\right)^{2.3}$$

$$= \underline{£27025}$$

(ii)

<u>Effect of variation of i on</u> ψ_{opt} and $C^{\mathrm{c}}_{\mathrm{opt}}$

$\underline{i = 0.08\ \mathrm{y}^{-1}}$ $\quad \therefore a^{\mathrm{c}}\ \dfrac{0.08(1.08)^{20}}{(1.08)^{20}-1} = 0.102\ \mathrm{y}^{-1}$

$$L = \frac{0.102 \times 1850}{0.07 \times 5000} = 0.538$$

$$\psi_{\mathrm{opt}} = \frac{1}{1 + (2.3 \times 0.538)^{\frac{1}{3.3}}} = \underline{0.484}$$

$$C^{\mathrm{c}}_{\mathrm{opt}} = 4400 + 17.2 \times 1850\left(\frac{0.484}{1-0.484}\right)^{2.3} = \underline{£31863}$$

$\underline{i = 0.16\ \mathrm{y}^{-1}}$ $\quad \therefore a^{\mathrm{c}} = \dfrac{0.16(1.16)^{20}}{(1.16)^{20}-1} = 0.169\ \mathrm{y}^{-1}$

$$L = \frac{0.169 * 1850}{0.07 \times 5000} = 0.892$$

$$\psi_{\mathrm{opt}} = \frac{1}{1 + (2.3 \times 0.892)^{\frac{1}{3.3}}} = \underline{0.446}$$

$$C^{\mathrm{c}}_{\mathrm{opt}} = 4400 + 17.2 \times 1850\left(\frac{0.446}{1-0.446}\right)^{2.3} = \underline{£19741}$$

<u>Effect of variation of</u> $C^{\varepsilon}_{\mathrm{IN}}$ on ψ_{opt} and $C^{\mathrm{c}}_{\mathrm{opt}}$

$c^{\varepsilon}_{\mathrm{IN}} = 0.05$ £/kWH

$$L = \frac{0.134 \times 1850}{0.05 \times 5000} = 0.992$$

$$\psi_{\mathrm{opt}} = \frac{1}{1 + (2.3 \times 0.992)^{\frac{1}{3.3}}} = \underline{0.438}$$

$$C^{\mathrm{c}}_{\mathrm{opt}} = 4400 + 17.2 \times 1850\left(\frac{0.438}{1-0.438}\right)^{2.3} = \underline{£22322}$$

$c_{\mathrm{IN}}^{\varepsilon} = 0.09$ £/kWh

$$L = \frac{0.134 \times 1850}{0.09 \times 5000} = 0.551$$

$$\psi_{\mathrm{opt}} = \frac{1}{1 + (2.3 \times 0.551)^{\frac{1}{3.3}}} = \underline{0.482}$$

$$C_{\mathrm{opt}}^{\mathrm{c}} = 4400 + 17.2 \times 1850 \left(\frac{0.482}{1 - 0.482} \right)^{2.3} = \underline{\pounds 31372}$$

Comment: A summary of the results of the second part of the problem is given in the table below.

$c_{\mathrm{IN}}^{\varepsilon}$ / £/kWh ↑			
0.09	-	$\psi = 0.482$ / $C^{\mathrm{c}} = \pounds 31372$	-
0.7	0.484 / £31863	0.463 / £27025	0.446 / £19741
0.05	-	0.438 / £22322	-
	0.08	0.12	0.16

$i/[\mathrm{y}^{-1}] \rightarrow$

As will be seen, low interest rate encourages the use of more expensive equipment which leads to higher rational efficiency. However, when the unit cost of input exergy increases, the lowest unit cost of the plant product is obtained by using a more expensive plant, which results in a higher process efficiency.

www.ingramcontent.com/pod-product-compliance
Lightning Source LLC
Chambersburg PA
CBHW051217200326
41519CB00025B/7147

* 9 7 8 1 7 8 2 2 2 0 0 0 8 *